WORKSHOP STATISTICS
Discovery with Data

Student Toolkit

Beth L. Chance
Allan J. Rossman

California Polytechnic State University
San Luis Obispo

Key College Publishing
Innovators in Higher Education

www.keycollege.com

Beth L. Chance and Allan J. Rossman
Department of Statistics
California Polytechnic State University
San Luis Obispo, CA 93407

Key College Publishing was founded in 1999 as a division of Key Curriculum Press® in cooperation with Springer New York, Inc. We publish innovative texts and courseware for the undergraduate curriculum in mathematics and statistics as well as mathematics and statistics education. For more information, visit us at www.keycollege.com.

Key College Publishing
1150 65th Street
Emeryville, CA 94608
(510) 595-7000
info@keycollege.com
www.keycollege.com

Published by Key College Publishing, an imprint of Key Curriculum Press.

Development Editor: Brendan Sanchez
Production Director: McKinley Williams
Production Project Manager: Beth Masse
Production Coordinator: Ken Wischmeyer
Copyeditor: Erin Milnes
Proofreader: Andrea Fox
Indexer: Victoria Baker
Art and Composition: Laura Murray Productions
Art and Design Coordinator, Cover Designer: Kavitha Becker
Cover Illustration: Roy Weimann
Printer: Von Hoffmann Press

Editorial Director: Richard J. Bonacci
General Manager: Mike Simpson
Publisher: Steven Rasmussen

Printed in the United States of America
10 9 8 7 6 5 4 3 2 1 06 05 04

ISBN 1-931914-67-2

We dedicate this *Toolkit* to our students—
past, present, and future.

Contents

To the Student		vii
To the Instructor		ix
Acknowledgments		x
UNIT I	**EXPLORING DATA: DISTRIBUTIONS**	
Topic 1:	Data and Variables	3
Topic 2:	Data, Variables, and Technology	7
Topic 3:	Displaying and Describing Distributions	11
Topic 4:	Measures of Center	17
Topic 5:	Measures of Spread	21
UNIT II	**EXPLORING DATA: COMPARISONS AND RELATIONSHIPS**	
Topic 6:	Comparing Distributions I: Quantitative Variables	29
Topic 7:	Comparing Distributions II: Categorical Variables	35
Topic 8:	Graphical Displays of Association	43
Topic 9:	Correlation Coefficient	47
Topic 10:	Least Squares Regression I	51
Topic 11:	Least Squares Regression II	63
UNIT III	**COLLECTING DATA**	
Topic 12:	Sampling	69
Topic 13:	Designing Studies	81
UNIT IV	**RANDOMNESS IN DATA**	
Topic 14:	Probability	91
Topic 15:	Normal Distributions	97
Topic 16:	Sampling Distributions I: Proportions	107
Topic 17:	Sampling Distributions II: Means	121
Topic 18:	Central Limit Theorem	131
UNIT V	**INFERENCE FROM DATA: PRINCIPLES**	
Topic 19:	Confidence Intervals I: Proportions	143
Topic 20:	Confidence Intervals II: Means	153
Topic 21:	Tests of Significance I: Proportions	161
Topic 22:	Tests of Significance II: Means	177
Topic 23:	More Inference Considerations	189
UNIT VI	**INFERENCE FROM DATA: COMPARISONS AND RELATIONSHIPS**	
Topic 24:	Comparing Two Proportions	195
Topic 25:	Comparing Two Means	203
Topic 26:	Inference for Two-Way Tables	211
Topic 27:	Inference for Correlation and Regression	217
INDEX		229

To the Student

This *Student Toolkit* serves as a supplement to the *Workshop Statistics* course books, and as such, contains additional resource material to aid your learning of fundamental statistical concepts, properties, and techniques. We hope that the additional examples, questions, and discussions that accompany each topic will enhance your discovery of the material in *Workshop Statistics*. We encourage you to work through the open-ended questions the way you would complete similar *Workshop Statistics* questions: apply your own knowledge first, then use the discussion in the *Toolkit* to test your understanding.

Most topics in this *Toolkit* consist of the following components:

- **Examples:** We begin each *Toolkit* topic with an example or two illustrating some of the primary ideas and methods of the corresponding *Workshop Statistics* topic. We have provided spaces following the examples so that you can write out your answers, applying what you learned in the main topic, before you read our explanations of the issues involved or check your work against the detailed solutions we've provided.
- **Key Concepts:** We highlight what we feel are the main ideas within each topic to help you organize and prioritize your knowledge and see the connections between topics.
- **Calculation Hints:** We offer advice for carrying out the calculations involved in the topic. We hope these hints will further facilitate your work with formulas so you can better focus on the underlying statistical concepts.
- **Common Oversights:** In our years of teaching with *Workshop Statistics*, we have seen students make certain mistakes with high frequency. We summarize some of these common errors in hopes of preventing you from falling into the same traps.
- **What Went Wrong?:** By presenting some erroneous solutions and asking you to identify and explain the errors, we give you additional opportunities in this section to test and clarify your knowledge. Our goal is to help improve your ability to evaluate statistical arguments and to anticipate common pitfalls. Such practice should enhance your own problem-solving skills and further develop your ability to decide if your own answers are reasonable.

- **Further Exploration:** In many topics we suggest some further activities or problems through which you can continue to explore statistical concepts. Many of these involve using Java applets that have been designed specifically to foster your explorations in a dynamic, visual environment. These applets can be downloaded free of charge at *www.rossmanchance.com/applets/*.

We hope you will find the discussions and advice in this *Student Toolkit* practical and supportive for improving your learning and achievement in this course.

Beth L. Chance and Allan J. Rossman

California Polytechnic State University
San Luis Obispo, California

To the Instructor

We strongly believe that students better acquire and retain knowledge they have constructed for themselves. Thus, we feel that the main strengths and unique aspects of *Workshop Statistics* are its discovery-based learning approach, the extensive use of real data, the focus on concepts, and the integration of technology. Yet we recognize that the discovery-based approach, when applied in its purest form, might not work for all students. Therefore, our goal in this *Student Toolkit* is to provide additional exposition that parallels *Workshop Statistics* in focus and notation, offering further support to students who need it, while remaining consistent with the strengths of the parent text.

Designed to supplement all five versions of *Workshop Statistics: Discovery with Data*, the *Toolkit* provides topic-specific support for students, regardless of which technology is being used. Each *Toolkit* topic opens with at least one **Example** that further elucidates concepts and methods developed within the complementary parent text topic. Just as in the parent text, space is allotted so that students can work out their answers on the page and reinforce concepts learned in the classroom. To further assist students, solutions—complete with explanations—are provided.

Every *Toolkit* topic then continues with three sections—**Key Concepts, Calculation Hints,** and **Common Oversights**—all of which endeavor to assist students in organizing and focusing their study of statistical concepts. Following these is **What Went Wrong?,** a section designed to enhance students' problem-solving skills by asking students to identify and explain errors present in flawed solutions. Finally, most topics conclude with a section titled **Further Exploration,** which offers additional activities for students to continue their investigation of difficult statistical concepts. In many instances, students use Java applets, enabling them to explore concepts in a dynamic, visual environment. These applets are available at no cost at *www.rossmanchance.com/applets/*.

Consistent with Workshop Philosophy?

Some might worry that a student guide such as this inhibits the role of student discovery that is central to *Workshop Statistics*, but we have constructed this *Toolkit* so that it enhances rather than diminishes student investigations. Students work through a set of examples in every topic, gaining practice in applying their knowledge and discovering where the holes in their understanding lie. As mentioned above, the What Went Wrong? section provides students with a way to construct their knowledge, allowing them to analyze errors commonly made by others so that those errors can be avoided. Finally, we believe that students will be in an excellent position to learn concepts presented in the *Toolkit*'s expository passages because the students already will have grappled with them in the corresponding *Workshop Statistics* activities.

We hope that you find this *Toolkit* to be a valuable resource for assisting your students' learning of statistics.

Acknowledgments

There are many to thank who have assisted in the development of the *Toolkit*. First, we gratefully acknowledge the following instructors and the valuable comments and suggestions they provided for earlier drafts of the *Toolkit:*

Judith Dill
Partners in Development Foundation, Honolulu, Hawaii

Johanna Hardin
Pomona College, Pomona, California

Tom O'Malley
Mount Union College, Alliance, Ohio

Stephanie Pepin
Montgomery College–Takoma Park, Montgomery County, Maryland

Sandra Perdue
University of Washington–Tacoma, Tacoma, Washington

William C. Rinaman
Le Moyne College, Syracuse, New York

Kim Robinson
Clayton College and State University, Morrow, Georgia

Allan Russell
Elon College, Elon, North Carolina

Robin Lock of St. Lawrence University, our co-author on the Fathom Dynamic Statistics™ version of *Workshop Statistics*, warrants special thanks for his extensive feedback. His review was extremely helpful in the development of the material.

We also wish to thank the many instructors who have shared with us their experiences of teaching with *Workshop Statistics*, as well as their advice on enhancing student learning. In addition, we gratefully acknowledge the invaluable guidance that we have received from our own students over the years as we have worked together toward the goal of learning about statistical concepts and methods.

Finally, we would like to thank Key College Publishing for supporting the development of the *Toolkit;* in particular, Mike Simpson and Richard Bonacci, who encouraged us to pursue this project, and Brendan Sanchez, who was very helpful and supportive in guiding it to fruition.

Unit I:

Exploring Data: Distributions

Topic 1: DATA AND VARIABLES

This topic focused on introducing you to some basic terminology that you will use throughout the book. The concept of a variable sounds simple, but you may actually need to practice a fair bit to be able to correctly identify variables. Recognizing variables is an important skill because you must identify the type of variable before you can determine which analysis to perform (for example, identifying the type of variable will tell you whether the correct graphical display is a bar graph or a dotplot). Another key idea in this topic—and throughout the book—is variability. A fundamental goal in statistics is to be able to describe and explain variability.

Example 1: The 1998 *Statistical Abstract of the United States* reports that a study found 1327 people who claimed to have done volunteer work during 1996, and 1392 who did not. Identify the observational units and the variable of interest. Suggest and construct an appropriate graphical display of these results, and describe what the graph reveals.

Example 2: A university professor asks all students in her class to estimate the distance from their hometown to the university. Identify the observational units and the variable of interest. Is the variable quantitative or categorical? Suggest a graphical display you could use if you had the data.

Example 3: A research firm claims that 80% of domestic cats live at least 10 years. Identify the observational units and the variable of interest. Is the variable quantitative or categorical?

Example 4: A study revealed that surgeons that performed a large number of surgeries had higher patient survival rates than surgeons that performed a lower number of surgeries. Identify the variable(s) of interest, and determine whether each is quantitative or categorical.

Example 5: Suppose a study is conducted to estimate the average size of households in your state. The researchers use your school and ask every person in the school how many people are in his or her family. They receive 200 responses and find the average to be 2.3 people. Identify the observational units and the variable of interest. Is the variable quantitative or categorical?

Solutions:

1. The observational units are people (a total of 2719). You can phrase the variable of interest as "Did the person do volunteer work?" where each person will give us a yes or no answer. Another phrasing is "Did you do volunteer work?" In this case, you have given the variable as a question you could pose to each observational unit. Doing this often helps you make sure that your statement of the variable truly varies from observational unit to observational unit. After defining this variable, you can see that it is binary and an appropriate graph would be a bar graph.

   As with most graphs of an individual variable, you should put the variable along the horizontal axis and convert the height of the bars to relative frequencies. From this graph you see that there was a slight majority of people in this sample who did not do volunteer work in 1996. [*Remember:* Always relate your discussion to the context of the problem.]

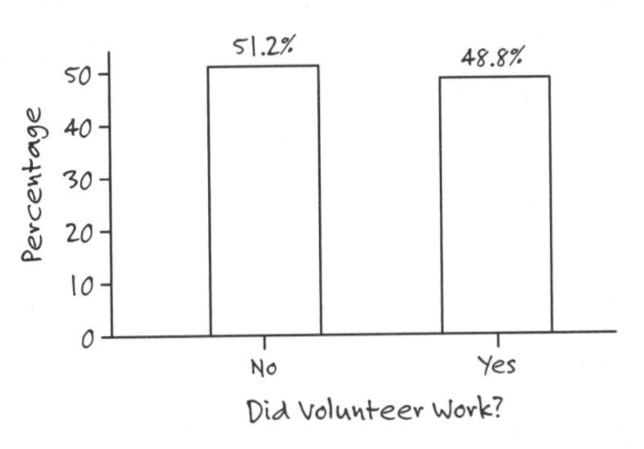

2. The observational units are students and the variable can be written as "distance from home" or "How far is your hometown from school?" Because the observation for each person is a number and it makes sense to discuss the "average number of miles," this is a quantitative variable. If you had the individual data values, then a dotplot would be an appropriate way to display the distribution. [*Remember:* Defining observational units first is very helpful in determining the variable.]

3. The observational units are domestic cats, but there are both numbers and categories floating around here. Is the variable quantitative or categorical? Remember to think about the question you would ask each observational unit. Each cat could have been asked at the end of its life, "Did you survive at least 10 years?" Each cat would then answer yes or no, so the variable of interest is a binary categorical variable. If, instead, for each cat the researchers had recorded the numerical result—"How long did you live?"—this variable would be quantitative. However, the researchers are reporting the percentage in a particular category, so they do not seem to have the individual values, and you can treat the variables as categorical. Although a percentage has already been calculated, that is just the numerical summary of all the responses, not a cat-to-cat result, and therefore not the variable.

4. There are two variables. Since you are making comparisons about the surgeons, the observational units are the surgeons. The first variable is whether or not the surgeon performed a large number of surgeries. Perhaps the researchers measured this value quantitatively, recording the number of surgeries, but it looks as though they categorized this variable into "high volume" and "not high volume." The second variable is the patient survival rate for the surgeon, which is quantitative.

5. The observational units can be specified in several ways; you just need to make sure you are consistent:

   Observational units = students in the school
   Observational units = households (slightly more appropriate)

   In either case, the variable is the size of the household or number of people *per household.*

   Note that the variable is *not* 2.3 people. This value is a summary, not a description of the measurement made household to household.

KEY CONCEPTS

- A variable is a characteristic that varies from observational unit to observational unit. A helpful way to think about defining the variable is to imagine you are entering each observational unit into a spreadsheet. Can you enter a different response next to each observational unit?
- Correctly identifying and specifying variables is a fundamental skill in statistical analysis.
- Identifying a variable as quantitative or categorical will help you determine which statistical tools to apply to the data.

COMMON OVERSIGHTS

- Confusing the variable and a numerical summary of the variable. In Example 2, the variable is not "average distance from home." In Example 3, the variable is not "number of cats that survived at least 10 years." These are individual numbers that summarize all the data collected, not a question whose response varies for each observational unit.
- Confusing the variable and a description of groups or specific observational units. In Example 4, you might have been tempted to identify the variables as "those with many surgeries" and "those with not as many surgeries," but this is specifying a group of observational units instead of describing what was measured about the observational units. The number of surgeries is what is changing from surgeon to surgeon, so the correct way to describe this variable is "how many surgeries performed?" (quantitative) or "performed high number of surgeries?" (categorical).
- Confusing the variable with the research question. In Example 4, you might have been tempted to use the question "Do busier surgeons have higher survival rates?" as your variable. This question has a yes or no answer, but it applies to the overall study, not to the observational units. In other words, it does not produce a different response from each surgeon.
- Confusing the variable with the result. For instance, in Example 4, you might have been tempted to list as one of your variables "Busier surgeons have higher survival rates." However, this statement is a summary of the result.

  It might be helpful to force yourself to always fill in the blanks in the following sentence:

  We are recording ______________ from ______________ to ______________.
  (variable) (observational units)

  For example: We are recording the survival rate from surgeon to surgeon.

- Confusing the objects being measured with the most basic objects in the study. For example, in Example 4, you might have been tempted to choose the patients as the observational unit and list the variable as whether or not each patient lives or dies. However, the data compared in this study are at the surgeon level. We are recording information about the surgeons themselves. If you think of the data in terms of a spreadsheet, each surgeon will be on a different row, and their responses to the questions "Do you perform a large number of surgeries?" and "What is your patient survival rate?" are the two variables listed in columns.

Topic 2:

DATA, VARIABLES, AND TECHNOLOGY

2

Although the focus of this topic was introducing you to using technology to analyze data, other important concepts were presented as well, such as the importance of considering ratios in addition to counts and considering the appropriateness of a variable in making decisions (e.g., average SAT score as a measure of a state's educational progress).

Example: The following table displays data on calories for several Chinese foods (from *Center for Science in the Public Interest*, tabulated by the *Philadelphia Inquirer* on March 23, 1995).

Dish	Calories	Dish	Calories
Egg roll (1 roll)	190	House lo mein (5 cups)	1059
Moo shu pork (4 pancakes)	1228	House fried rice (4 cups)	1484
Kung pao chicken (5 cups)	1620	Chicken chow mein (5 cups)	1005
Sweet and sour pork (4 cups)	1613	Hunan tofu (4 cups)	907
Beef with broccoli (4 cups)	1175	Shrimp in garlic sauce (3 cups)	945
General Tso's chicken (5 cups)	1597	Stir-fried vegetables (4 cups)	746
Orange (crispy) beef (4 cups)	1766	Szechuan shrimp (4 cups)	927
Hot and sour soup (1 cup)	112		

a. Identify the observational units and variable. Is the variable quantitative or categorical?

b. Produce an appropriate graph and informally describe the distribution of calories in these foods.

c. Do you think it is reasonable to use these data to rank the foods from least to most in terms of caloric content? Explain how else you might look at the data if you were interested in counting calories.

Solution:

a. The observational units are the Chinese food dishes, and the variable is the number of calories in the dish, which is quantitative.

b. Here is a dotplot:

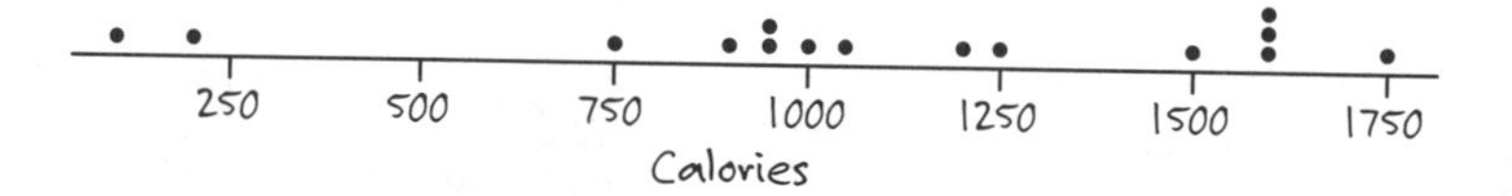

The calorie amounts appear to be centered around 1000 calories but range from about 100 to about 1750 calories. You can especially see that two dishes have much lower calorie amounts than the other dishes. Returning to the data table, you see these are the egg roll and the hot and sour soup.

c. The outliers suggest that you might want to make some adjustment for the serving size. It is a little unfair to compare the egg roll or the hot and sour soup to the main dishes because one egg roll or one cup of soup does not make a meal, whereas 4 cups of sweet and sour pork would be quite satisfying. Thus, you could convert all of the calorie amounts to a rate per cup or to some measure of "one serving" for a fairer comparison among the foods.

KEY CONCEPTS

- Computer software can be very helpful in exploring and analyzing data sets. It is worth your time early in the course to become comfortable with a statistical software package or a graphing calculator. Such technology will save you increasing amounts of time as the course progresses. You might want to make a list for yourself of all the computer or calculation functions that you learn as you go along for easier reference later in the course.

- Keep in mind that graphs and calculations are only the first step. Once you complete your graphs, you will have to interpret the information, make comparisons and conclusions (in context), and then support your statements with the calculations and graphs.

- You will see many reports where a count is given but a rate or proportion would be much more informative. Sports data is a prime example. You may hear that Joe

Quarterback completed 11 passes in a particular situation—but maybe he attempted 2000! Or Dr. Dentist's patients have only had 1 cavity—but maybe they have already lost most of their teeth! More seriously, comparing Rhode Island to New York by looking at raw numbers, such as number of college graduates, would be silly because New York has many more people than Rhode Island. A more useful comparison would be comparing the rates of college graduates in the two states, where those rates are calculated by dividing the number of college graduates by the state population. In short, you should always consider the number of observations when evaluating data.

COMMON OVERSIGHTS

- Confusing *percentage* and *proportion*. A percentage is a number between 0% and 100% (inclusive), whereas a proportion is a number between 0 and 1. Multiplying a proportion by 100 converts it to a percentage.

Topic 3:
DISPLAYING AND DESCRIBING DISTRIBUTIONS

3

This topic expanded your tools for displaying and discussing distributions of data, specifically quantitative data. You became familiar with two new graphical displays (stemplots and histograms) and developed a checklist of features to comment on when describing a distribution. You will use these tools throughout the course, so it's important to fully understand and effectively discuss the information conveyed by these graphs. In later topics you will come across more complicated situations, but the graphical representations and descriptions will be the same. In particular, you should quickly get in the habit of discussing shape, center, and spread when describing the distribution of a quantitative variable.

Example: The following dotplot and histogram present the data about calories in Chinese food discussed in *Toolkit (TK)* Topic 2.

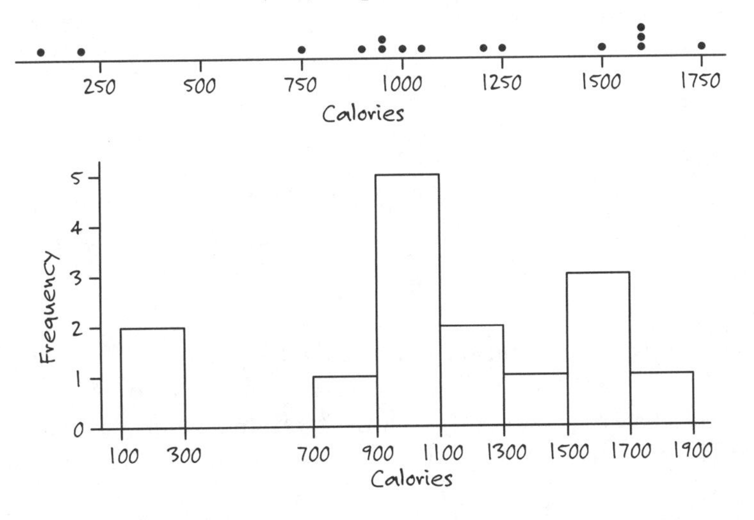

Describe the distribution of the number of calories in these foods.

Solution:

If there are any outliers, it's a good idea to consider them first. Then describe the shape, center, and spread.

Outliers

You saw earlier that the hot and sour soup and the egg roll have much lower calorie counts than the other foods. An explanation for this is that they are appetizers and the values given here don't represent a complete serving size. Given that they represent a different observational unit, you may feel justified in removing them from the analysis.

Shape

It is difficult to discern an overall shape from these graphs because the number of values is small and the dotplot spreads the information out over a large range. Even in the histogram it is difficult to see an overall pattern. You may want to ignore the outliers and describe the distribution for the remaining foods. Reconstructing the dotplot and histogram gives the following:

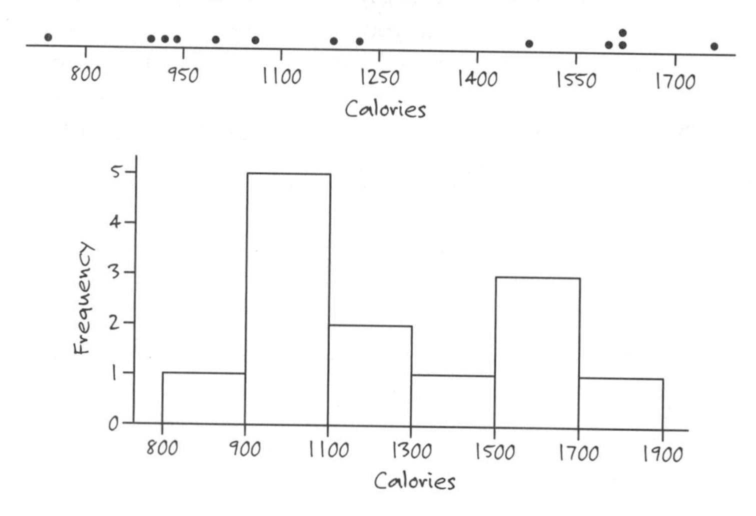

From the dotplot, you would say that the distribution is fairly "even," that is, the data points are evenly spread out. From the histogram you would say that the distribution is slightly skewed to the right, because the right side of the distribution has the longer tail. It is a good habit to look at more than one plot when exploring your data.

Center

If you ignore the outliers, you would say that 1200 calories is a fairly central value or a typical calorie count.

Spread

If you again ignore the outliers, you would say the distribution of calories has a large spread, ranging from 746 to 1766 calories. The calorie counts of these foods are not very consistent.

When discussing shape, center, and spread, it is up to you to decide whether to include the outliers. Keep in mind that the goal is to describe the bulk of the data, the general tendencies and patterns. Therefore, you don't want a few unusual observations to color your overall impression of the distribution too much.

KEY CONCEPTS

- In describing a distribution, first consider outliers. If you find any, make sure that you return to the data set and identify the individual observations. You should investigate any possible explanations for the outliers. In particular, make sure they do not arise from any typographical errors.
- The three main features to *always* consider in describing a distribution are the shape, center, and spread of the distribution. It is often useful to trace over the graph with your pencil to help you see the overall shape.
- Don't forget to comment on any other unusual features, such as granularity, if they occur in the distribution.
- It is very important that you always label and scale the horizontal axis of a histogram or a dotplot. Think again about the observational units and the variable being measured on them. Except for stemplots, the convention is to put the variable along the horizontal axis and the number or proportion of observations that occur at each value or in each subinterval on the vertical axis.
- Another important habit is looking at several different types of graphs. Don't stop after the first one. There are advantages and disadvantages to each type of graph, and each will enhance your understanding of the data. The stemplot enables you to see each individual numerical value but is less practical with a large data set. Histograms are helpful in that they group the data to help you focus on the overall pattern, but you lose some detail as a result. For example, in the last histogram in the example, you see how many foods have fewer than 1100 calories (a bar edge) but not how many are have fewer than 1200 since that value lies within a bar.

COMMON OVERSIGHTS

- Confusing "skewed to the left" and "skewed to the right." Keep in mind that the direction of the skew is indicated by the longer tail. Some students like to remember this by noting that "skewed to the right" means "piled up on the left," and vice versa.
- Not relating your comments to the context of the data. It is not sufficient to give numbers and say "skewed to the right," for example, without clarifying the variable that is being measured and the units of measurements (e.g., calories). After you have written your description of a distribution, stop and ask yourself whether someone who reads your description alone would understand whether you are talking about calories of foods or weights of animals or weekly incomes of employees.
- Using the word "it" when the antecedent is unclear. Do not say, "It is skewed to the right." Rather, say, "The distribution of calories in these foods is skewed to the right."
- In constructing a stemplot, not ordering the leaves within each stem or not including all stems, even those that do not have leaves. Also make sure you record a leaf in the stemplot for each observation; even if the leaf value has already been used you should repeat the value as many times as it occurs in the data set.
- Not labeling and scaling the horizontal axis.

- Using the word "even" when you mean "symmetric." The word "even" is reserved for talking about distributions that are fairly flat, that is, distributions in which the heights of the bars are similar. If you mean that the left half of the distribution looks similar to the right half, you should use the word "symmetric."

WHAT WENT WRONG?

The following are examples of errors in the analysis of an example problem. In each case, explain what is wrong. (We strongly suggest that you try to figure these out for yourself first before looking at the solutions.)

1. For the Chinese food example, suppose a student gave the following graph and then said that the distribution was fairly even, with Hunan tofu being a "middle value" of the distribution. How would you respond?

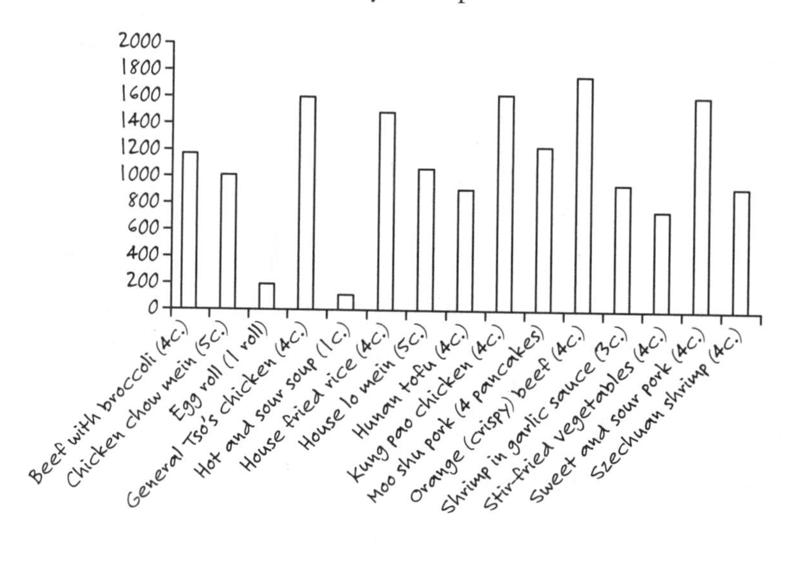

2. For the Chinese food example, suppose a student gave the graph at the top of the next page and then said that the distribution was symmetric, with orange beef having a typical amount of calories. How would you respond?

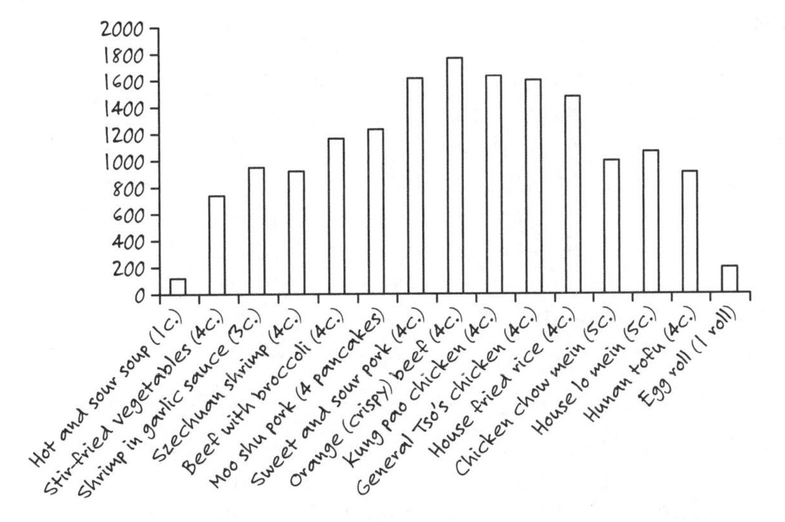

3. Suppose that a student produces the following stemplot and verbal description of the distances (in miles) of the 72 hikes described in the book *Day Hikes in San Luis Obispo County* by Robert Stone (Day Hike Books, 2000):

Stem	Leaves
0	68
1	00000555555588
2	0000000122555556668
3	000002224458
4	00000566
5	55668000
6	0000
7	004
9	5

leaf unit = 0.10

Verbal description: The data are skewed to the left. Most are at the value 20. The center is in the 20's. The spread is from 6 to 95. There are no outliers, and there is no granularity.

What is your critique of this analysis?

Solutions:

1. This choice of graph is problematic because it focuses on the individual foods instead of the distribution of the number of calories. The observations have been ordered alphabetically and do not allow us to easily discern the shape, center, and spread of the distribution. You could try to approximate the average height of these bars to determine the center of the distribution, but even that is more difficult here than with a proper histogram. Although this graph displays bars, it is not a histogram. You need to make sure the variable of interest is along the horizontal axis; then the heights of the bars will correspond to the number of observational units at each value or within each range of values.

2. This distribution looks "prettier" than the previous one, and the shape is indeed symmetric, but again the graph focuses on the individual foods rather than the calorie counts. The orange beef has the largest number of calories, and although you can tell this quickly from the graph, it is difficult to describe the pattern of the distribution of the calorie values. In fact, you saw in the previous discussion of shape in the topic that the distribution of calories is not symmetric, so this graph is very misleading.

3. There are two errors in the construction of the stemplot. The leaves on the 5 stem are not arranged in order (they should be listed as 00055668), and the 8 stem should be inserted in the graph even though there are no leaves for that stem. There are many problems with the verbal description. First, the leaf unit is 0.10, so these distances are between 0.6 and 9.5 miles, not between 6 and 95 miles as the description claims. The center is therefore in the 2's, not in the 20's. The shape of the distribution of hike distances is skewed to the right, not to the left, because the longer tail is at the high end. Although 2.0 miles is the mode of the distances, it is not true that most hikes are 2.0 miles. The 9.5 mile hike is an outlier, although that is hidden by the missing 8 stem in the plot. There is also a type of granularity in that almost half of the hikes are an integer number of miles; in other words, their leaf is 0. Finally, one important and glaring problem with the verbal description is that it makes no reference to the context. From this description, the reader has no idea whether the data are distances of hikes or scores on an exam or hours of sleep in a week or . . . . The description should not only mention that the data are hike distances but should identify the measurement units as miles.

Topic 4:
MEASURES OF CENTER

The previous topic discussed the center of a distribution very informally. In this topic, you worked with the mean and median as numerical measures of center. These numbers often come out differently for the same data set, and it is important to consider the information conveyed by each.

Example:

a. For the Chinese food study, how do you think the mean and median number of calories compare?

b. What if the two outliers are removed?

Solution:

a. The full data set has two low outliers. These two outliers could be low enough to pull the mean down lower than the median.

b. When the two outliers are removed, the distribution appears slightly skewed to the right, as you saw in the previous topic. This skew would lead you to suspect that the mean will be larger than the median when the outliers are removed.

With all 15 food items: mean = 1092 calories; median = 1059 calories
With 13 food items: mean = 1236.3 calories; median = 1175 calories

The first prediction is incorrect, but the second prediction is correct. Although you can make conjectures based on the shape of the data, you still need to calculate the values. When estimating the mean, also keep in mind that it serves as the balance point of the distribution. Think of each data point as a weight lying

4

on a board, and try to balance the dotplot on a fulcrum. Where would you place that fulcrum?

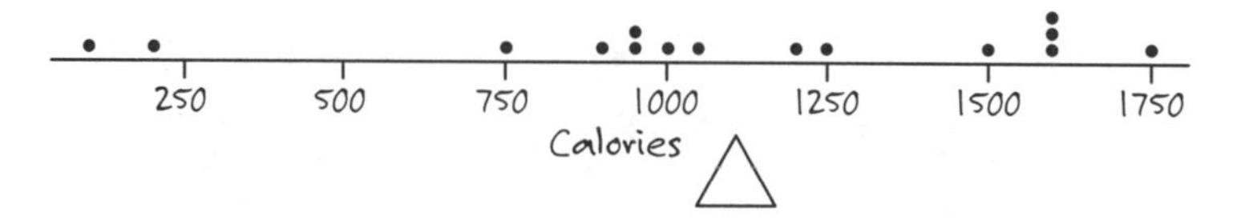

The mean would be pulled to the left by the outliers, but some food items have large numbers of calories that are as far from the center as the outliers. Then there are a few more foods around 1250 calories and one food around 750 calories. Because there are a few more large numbers, which are not balanced by lower values on the left, the mean is slightly higher than the median.

When the outliers are removed, the distribution is somewhat skewed to the right and your prediction that the mean is greater than the median holds true.

KEY CONCEPTS

- The mean, which is calculated using all of the numerical values in the data set, is not *resistant* to the effect of outliers. The median, which uses only the middle value or the middle pair of values, is resistant to the effect of outliers.
- When the *location* of the median values falls between two data values, take their average as the *value* of the median.
- Keep in mind that the mean and median only describe one aspect of the distribution: center. They don't give you all of the information about a distribution; for example, they say nothing about the variability in the data. There may be situations where a single number is insufficient to capture the typical values of a distribution (e.g., a bimodal distribution is better described by two numbers).
- Although the median can be interpreted as a typical value, the mean is the balance point and always tells you the average of all the numerical values. You can report both but should realize they measure slightly different things. Or you may choose to report only one, based on the research question and/or the shape of the data.
- Keep in mind that variability is a key concept in statistics. You should try to think about why values vary for different situations and try to anticipate variable behavior. Ask yourself, for example, "Will the distribution be skewed or symmetric?"

CALCULATION HINTS

- When reporting the mean and median, always ask yourself whether the values seem plausible considering the context. For example, if you calculate the mean hours of sleep last night for a student in your class to be 19.6, you should recognize that you've probably made a calculation error. In contexts with which you are less familiar, look back at a graph of the distribution to see whether the mean and median values make sense.

COMMON OVERSIGHTS

- Not sorting the data before calculating the median.
- Reporting the result of the calculation $(n + 1)/2$ as the value of the median instead of realizing that this indicates the *location* of the median.
- Incorrectly using the word "most" with regard to the mode. The mode indicates the most frequently occurring value, but that is not the same as saying that "most of the observations" are at that value. For example, in *Workshop Statistics (WS)* Activity 4-4, the modal pamphlet readability level was 8. However, it would be incorrect to say, "Most pamphlets had a readability level of 8;" since $8/30 = .27$, only 27% of the pamphlets had that value.
- Not thinking about the effect of unusual or outlying observations on your calculations.
- Trying to calculate a mean or median with categorical data. Keep in mind that these measures of center apply to quantitative variables only.

WHAT WENT WRONG?

The following are examples of errors in the analysis of the example problem. In each case, explain what is wrong.

Data were obtained for all U.S. roller coasters. One variable measured was the number of inversions made by the coaster during one run. The following table displays the results for 144 coasters.

Number of inversions	0	1	2	3	4	5	6	7
Tally (count)	84	7	12	11	9	9	3	9

1. A fellow student reports the median to be 3.5.

2. A fellow student reports the mean to be 18.

3. A fellow student reports the median to be 9.

4. A fellow student reports the median to be 0.

Solutions:

1. This response can arise by treating the eight categories as eight observations and applying the $(n + 1)/2$ rule to find that the location of the median is between the 4th and 5th observations. However, n in this formula is the total number of observations in the data set, not the number of distinct results. The observational units are the roller coasters, so $n = 144$. The median, therefore, occurs between the 72nd and 73rd observations. You can use the tally information to determine where this is. Since 84 of the coasters had 0 inversions, the 72nd and 73rd values are also equal to 0, so the median number of inversions is 0. The median would have been 3.5 in this case if there were an equal number of roller coasters at each *number of inversions* value.

2. This response can not be correct: no roller coaster in this data set has more than 8 inversions, so the mean number of inversions cannot be 18. The number 18 is the mean of the tallies, not of the number of inversions. In other words, this response treated the eight categories, rather than the roller coasters, as the observational units. To determine the mean, we would add 84 zeros to 7 ones to 12 twos to 11 threes to 9 fours to 9 fives to 3 sixes to 9 sevens and then divide by the total number of roller coasters, 144:

$$\frac{84(0) + 7(1) + 12(2) + 11(3) + 9(4) + 9(5) + 3(6) + 9(7)}{144} = 1.57 \text{ inversions}$$

   Notice that this number makes much more sense. Most of the coasters have 0–3 inversions, so the mean should be around 1 or 2.

3. This response also treated the categories rather than the roller coasters as the observational units and found the median of the counts 3, 7, 9, 9, 9, 11, 12, and 84. Once again this answer is clearly impossible, because none of these coasters has as many as 9 inversions.

4. The median is indeed zero, as shown in the solution to part a, but this solution does not put this figure in context—Zero what?

   Note: It is easier to see that parts 1–3 are wrong after you have seen the graph:

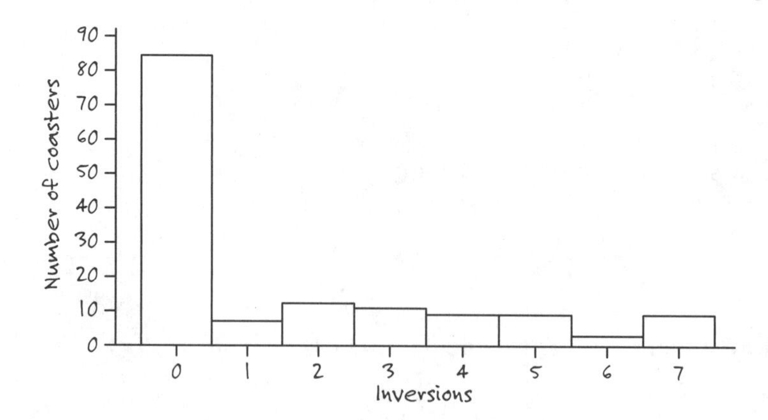

   Either measure of center should be low, around 0 or 1, with the mean slightly higher than the median due to the long right tail of the distribution.

Topic 5:
MEASURES OF SPREAD

Just as *WS* Topic 4 gave different ways to numerically measure the center of a distribution, *WS* Topic 5 suggested several ways to measure the spread of a distribution—range, interquartile range, and standard deviation. You saw that the range and standard deviation are not resistant and thus should be used with caution if there are extreme outliers in the distribution. Similarly, the interquartile range is more appropriate for a skewed distribution than the standard deviation. Thus, you need to look at the graph, and the research question of interest, to decide which numerical summary to report. In return, the impressions you glean from graphs should always be reinforced by the numerical summaries. From now on, you will want to look at both graphical and numerical summaries in describing your data.

Calculating the standard deviation is rather complex and we hope you will begin to rely on technology more and more to carry out these calculations. Still, the most important focus of this topic is the reinforcement of the meaning of "variability" (*WS* Activity 5-7). You should also be aware of what values are possible for the standard deviation (e.g., if you get a negative standard deviation, something has gone terribly wrong).

In this topic, you also learned a different interpretation of standard deviation through the empirical rule and a strategy for determining the relative position of an observation within a data set through the *z*-score.

Example 1: For the Chinese food example, report the five-number summary and the interquartile range for the calorie counts.

5

Example 2: The following boxplots display the distributions of speed (in miles per hour) for steel and for wooden roller coasters in the United States:

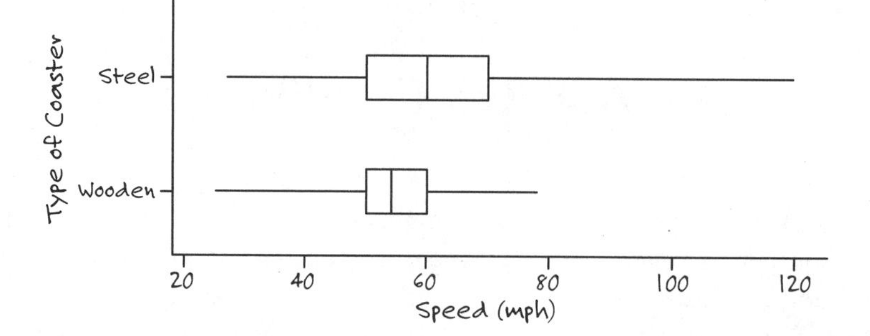

a. Do these boxplots enable you to determine whether there are more wooden or steel roller coasters? Explain your answer.

b. Do these boxplots enable you to say which type of coaster has a higher percentage that go faster than 60 mph? If so, what percentage of each type goes faster than 60 mph? Explain your answer.

c. Do these boxplots enable you to say which type of coaster has a higher percentage that go faster than 50 mph? If so, what are the percentages? Explain your answer.

d. Do these boxplots enable you to say which type of coaster has a higher percentage that go faster than 40 mph? If so, what are the percentages? Explain your answer.

Solutions:

1. Here is a stemplot for the data:

```
0 | 11
0 | 7999          leaf unit = 100
1 | 00124
1 | 5667
```

Because there are 15 observations, you found previously that the median is located at the eighth observation (1059 calories). This leaves seven food items below the median and seven food items above the median. With seven values, the median falls at the fourth observation. So Q_1 is equal to the fourth observation from the bottom, and Q_3 is equal to the fourth observation from the top. Although you can circle these on the stemplot (0|9 and 1|5), it is still important to go back to the data table to see the full numerical values: 907 and 1597. Similarly, to determine the minimum and maximum, return to the original data values: 112 and 1766 calories. So the five-number summary is:

Min	Q_1	Median	Q_3	Max
112	907	1059	1597	1766

The interquartile range is then calculated to be 1597 − 907 = 690 calories.

You saw in earlier topics that two of the foods had unusually low calorie counts. Notice that even if these two foods both had zero calories, the numerical value of the interquartile range would not change. However, the numerical value of the standard deviation would be much larger.

2. **a.** No. These boxplots do not reveal anything about the sample sizes involved.

   b. Yes. The median speed for the steel coasters is 60 mph, and the upper quartile of the wood coaster speeds is 60 mph, so you know that 50% of steel coasters exceed 60 mph and only 25% of wood coasters exceed 60 mph.

   c. Both types have 50 mph as the lower quartile of their speeds, so for both types 75% exceed 50 mph.

   d. No. The value 40 mph falls between the minimum and the lower quartile for both types of coaster, but you can't say how the other observations in that region are distributed and what percentage fall above 40 mph. Therefore you can't say which type has a greater percentage of coasters that exceed 40 mph.

KEY CONCEPTS

- The variability or spread (we use these words interchangeably) of a distribution refers to how far apart the data values tend to be from the middle of the distribution. In fact, you can loosely interpret standard deviation as a typical or average deviation from the mean. Similarly, the interquartile range gives information about the "width" of the middle 50% of the distribution.

- Variability/spread is different from "bumpiness" or variety, as you saw in *WS* Activity 5-7.
- In describing the amount of variability in a distribution, try to focus on the bulk of the distribution, not just a few extreme values.
- The mean and standard deviation are best used with symmetric data, whereas the five-number summary, including the interquartile range, is always appropriate (for data sets with at least five values), particularly for skewed distributions.
- The z-score, or standardized score, is an important "ruler" that statisticians use for measuring distance with symmetric distributions. It provides a very useful reference for where an observation falls in a distribution because it takes into account both the center and the spread of the distribution.

CALCULATION HINTS

- When determining the quartiles, if there is an odd number of observations, so that the location of the median is at a particular value, the convention is not to include the median in the lower half and the upper half of the data. In the previous example there were 15 observations but you used only 7 in each of the data "halves."
- Keep in mind that the range and interquartile range are each reported as one number. For example, instead of saying that the range of the temperatures in San Francisco is 49 to 65 degrees, calculate the range to be $65 - 49 = 16$ degrees. Similarly, the IQR is 10 degrees, not 52.5 to 62.5 degrees.
- Remember to keep track of the measurement units and to report them with your calculations.
- Don't be alarmed if the calculations you generate by hand do not match the values reported by the computer or calculator, particularly for quartiles, because the technology may use a slightly different algorithm.
- When performing calculations by hand, try to maintain as many digits of accuracy as possible in intermediate calculations and round only the final answer. One or two decimal places are usually plenty for the final answer.

COMMON OVERSIGHTS

- Confusing variability with "bumpiness" or variety. Even if you can order the distributions in *WS* Activity 5-7 correctly now, confusing variability with variety is a bad habit that's easy to fall into. You should check your interpretation of variability frequently throughout the course. Keep in mind that the only way to have no variability is for all of the values to be identical ($s = \text{IQR} = 0$).
- Applying the empirical rule even if your data are not symmetric (or not checking first whether your data are symmetric).
- Trying to calculate the numerical summaries on categorical data or on the tallies instead of the data values.

- When constructing a boxplot, placing the line in the box at the mean instead of the median.
- Believing that the five-number summary reveals everything about a distribution. Remember that any summary loses some of the information present in the original data. It's possible for two distributions to have identical five-number summaries and yet have some different features.

WHAT WENT WRONG?

The following are examples of errors in the analysis of the example problem. In each case, explain what is wrong.

Suppose that students are asked to make up a data set of ten quiz scores on a 1–5 scale. First they are asked to create a data set with the largest amount of variability possible. Then they are asked to create a data set with the least amount of variability possible. Explain the error in the student's thinking for each of the following responses:

1. Most variability: 1, 1, 2, 2, 3, 3, 4, 4, 5, 5, as shown in the following histogram:

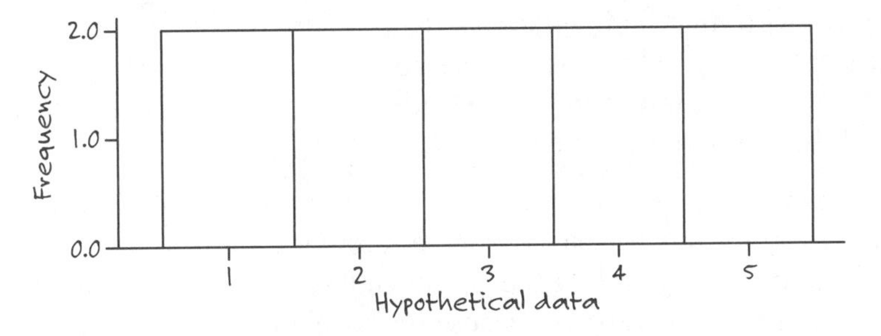

2. Least variability: 1, 1, 2, 2, 3, 3, 4, 4, 5, 5, as shown in the preceding histogram.

Solutions:

1. The student was probably thinking that having every possible value (1–5) represented gave this data set the most variability possible. But this is not correct, because the most variability is achieved by putting the values as far away from the center as possible. The correct answer is 1, 1, 1, 1, 1, 5, 5, 5, 5, 5. This puts the center at 3 (as measured by mean or median) and all of the values as far from the center as possible, as shown in the following histogram:

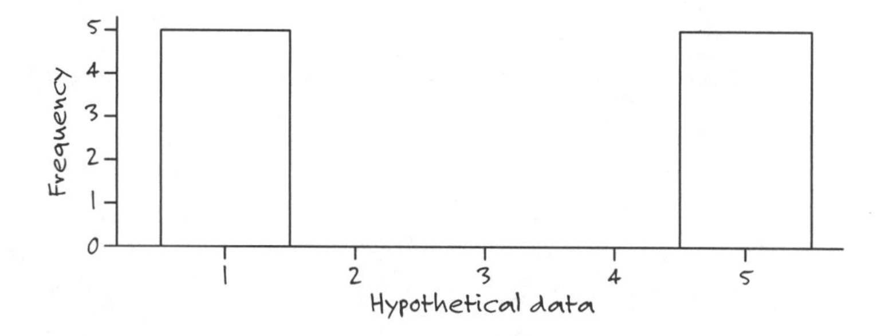

The IQR for this example is 5, and the standard deviation is 2.11. Note that this is consistent with the rough guide that the standard deviation measures how far from the mean the data values are on average, because in this example all of the data values are 2 units away from the mean. For the first data set suggested, the IQR is 2.5 and the standard deviation is 1.49.

2. The student was probably thinking that the example proposed has no variability because every rating appears the same number of times. Although it is true that there is no variability among the counts (tallies) in this example, there is definitely variability among the ratings because they are not all the same. A correct example showing the least variability possible would put all of the ratings on the same value, which could be any of the numbers from 1 to 5. In this case the standard deviation and IQR would both equal zero.

Unit II:

Exploring Data: Comparisons and Relationships

Topic 6:

COMPARING DISTRIBUTIONS I: QUANTITATIVE VARIABLES

In this topic you combined the tools you learned in the earlier topics with the goal of comparing the distribution of a quantitative variable between two or more groups. You also learned how to construct modified boxplots, which included a specific criterion for determining which observations are far enough from the rest of the data to be considered outliers.

Example: The Survey of Study Habits and Attitudes (SSHA) is a psychological test that measures the motivation, attitude toward school, and study habits of students. Scores range from 0 to 200, with higher values indicating more positive attitudes and habits. A selective private college gave the SSHA to both female and male first-year students, with the following results:

Women:
154 109 137 115 152 140 154 178 101 103 126 126 137 165
165 129 200 148

Men:
108 140 114 91 180 115 126 92 169 146 109 132 75 88 113
151 70 115 187 104

a. What variable is being measured for these women and men? What type of variable is this?

b. Produce appropriate numerical and graphical summaries to compare the distributions of SSHA scores between these women and men.

c. Write a paragraph comparing these two distributions.

Solution:

a. The variable is the SSHA score, which is quantitative.

b. You can construct stacked dotplots and/or boxplots of the two distributions of the same scale.

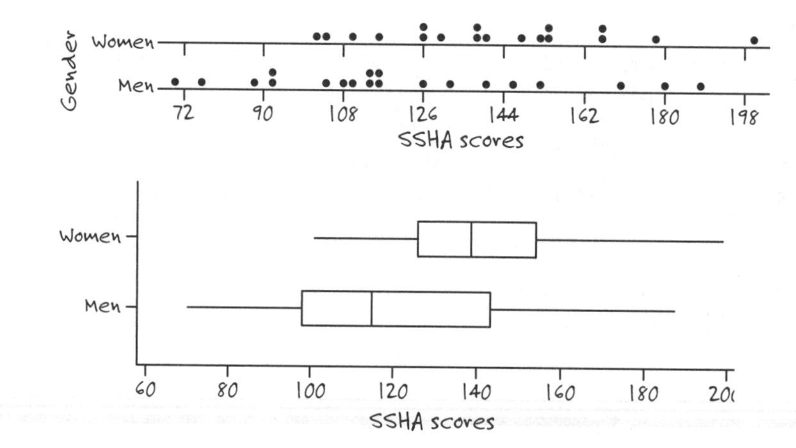

The five-number summary for the SSHA scores for each group is shown in the following table:

	Min	Q_1	Median	Q_3	Max
Women	101	126	138.5	154	200
Men	70	98	114.5	143	187

[*Note:* If you did these calculations in Minitab®, you will get slightly different values for the quartiles (women: 123.25 and 156.75; men: 95 and 144.50).]

c. You can also see that the SSHA scores of women tend to be higher than the scores of men (median 138.5 versus 114.5; in fact, all five numbers in the five-number summary are larger for women). However, the male scores are a bit more spread out (IQR 45 versus 28). Although it is hard to judge shape from boxplots, it appears that both groups' distributions of SSHA scores are slightly skewed to the right (judging from the right sides of the boxes being a bit longer than the left sides and the right whiskers being longer than the left whiskers).

> There are many ways to write a good paragraph. You should demonstrate that you have clearly thought about the data, have commented on at least shape, center, and spread, and have related your comments to the context of the study.

KEY CONCEPTS

- A statistical tendency is a general trend but not a hard-and-fast rule. We all know that men tend to be taller than women. Put another way, men are taller than women on average, or most men are taller than most women. But lots of individual women are taller than lots of individual men. Similarly, in the preceding example, women tend to score higher than men on the SSHA test, but this not to say that every woman scored higher than every man.
- When comparing distributions you should always discuss the three most important features—shape, center, and spread—and then any other features, such as outliers, that are present. As always, remember to relate your comments to the context, so that someone reading only your description would know whether you are comparing SSHA scores between men and women or bowling scores between children and adults.
- Boxplots are very useful for comparing groups, but you might also want to use stemplots or other graphs to see whether there are any other unusual aspects to the shapes of the distributions. In the SSHA scores example, the shapes of the distributions for women and men are similar and fairly symmetric, judging from the following dotplots, but you should always check another type of plot rather than assume there are no clusters or gaps.

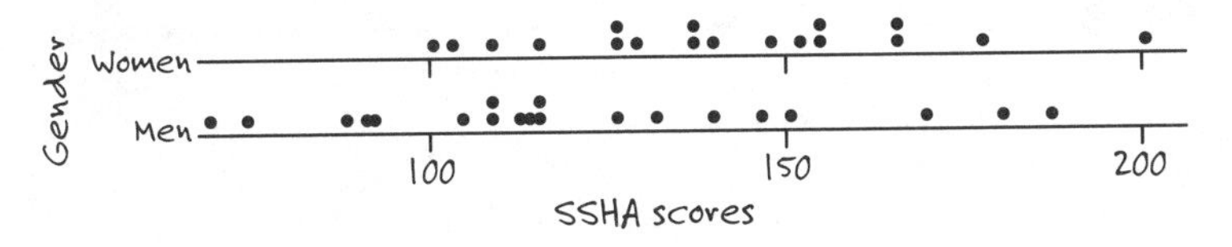

- Modified boxplots are very useful for determining numerically whether observations are extremely large or extremely small. Keep in mind that observations may be unusual in other ways as well (e.g., being the only noninteger).

6

COMMON OVERSIGHTS

- When constructing a modified boxplot, incorrectly extending the whisker to the cut-off point ($Q_3 + 1.5 \times IQR$ or $Q_1 - 1.5 \times IQR$) rather than to the next observation in the data set that lies within these cut-offs. The endpoints of the whiskers tell you about specific values in the data set.
- Mistakenly using the mean or median instead of the quartiles to detemine outliers. There are other outlier "rules" that use the mean and standard deviation (e.g., that any data point more than three standard deviations from the mean is an outlier). However, it is important not to mix and match these rules (e.g., "more than $3 \times$ IQR from the mean" or "$1.5 \times$ IQR from the mean") or extend the cut-offs from the median instead of the quartiles (median $\pm 1.5 \times$ IQR).
- Checking for outliers in only one direction. Be sure you check on both the high and the low ends.
- Stating conclusions without providing supporting evidence. For example, don't simply say that women tend to have higher SSHA scores than men. You can partially support this statement by pointing out that the median SSHA score is higher for women than for men by 24 points (138.5 to 114.5). You can provide further support by pointing out that all values in the five-number summary are higher for women than for men. In fact, looking at those summaries reveals that at the higher end of scores, men and women come closer than at the lower end.

WHAT WENT WRONG?

The following are examples of errors in the analysis of the example problem. In each case, explain what is wrong.

1. The following boxplots present the lengths (in feet) of roller coasters, classified by type of coaster (steel or wooden):

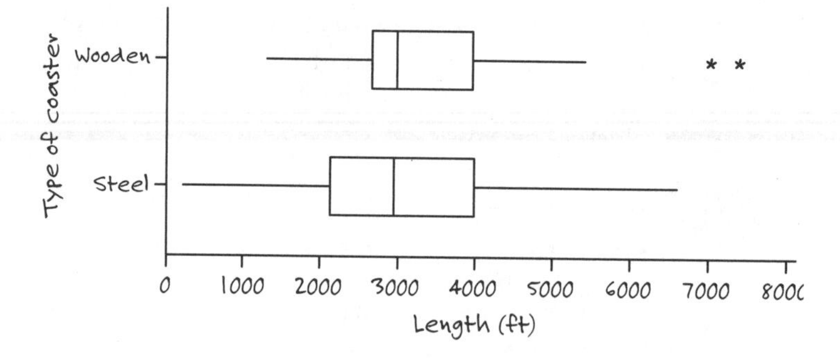

Suppose that a student's commentary on this graph is, "The medians both equal 3000, so there is no difference between the lengths of steel and wooden roller coasters." Identify several problems with this analysis.

2. The following stemplot present the distances (in miles) of the 72 hikes described in the book *Day Hikes in San Luis Obispo County*:

Stem	Leaves	
0	68	leaf unit = 0.1
1	00000555555588	
2	0000000122555556668	
3	000002224458	
4	00000566	
5	00055668	
6	0000	
7	004	
8		
9	5	

Suppose that a student produces this boxplot of the hike distances. Can you spot the error made in the boxplot? [*Hint:* You are welcome to calculate the five-number summary and conduct the outlier test, but you should be able to spot the error without doing those calculations.]

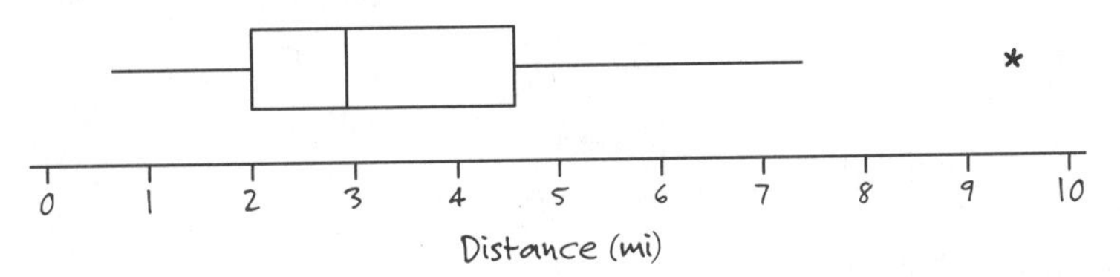

Solutions:

1. The medians represent only one feature (center) of a distribution. There is much more to comment on, and the wooden and steel coasters do differ with regard to lengths in several ways. There are two large outliers among the wooden coasters, which are longer than any of the steel coasters. Except for those two outliers, though, the wooden coasters have less variability in their lengths than the steel coasters do.

2. The right whisker extends too far. With the 9.5-mile hike identified as an outlier marked by an asterisk (*), the right whisker should extend only to the largest non-outlier, which is the 7.4-mile hike. The student has probably extended the whisker to the cut-off point for detecting outliers, rather than to the largest non-outlier in the data.

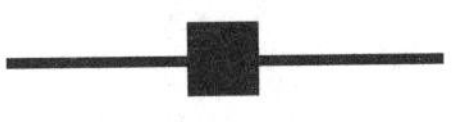

Topic 7:

COMPARING DISTRIBUTIONS II: CATEGORICAL VARIABLES

WS Topic 7 paralleled the previous topic in its treatment of comparing groups, but instead of making the comparison on a quantitative variable, you compared the groups on a categorical variable. After you identify your observational units and variables, if you determine that both variables are categorical, you will use the numerical and graphical methods from this topic to begin your analysis.

Example: Toward the end of 2003, there were many warnings that the flu season would be especially severe. Many more people chose to obtain a flu vaccine than in recent years. In January 2004, the Centers for Disease Control and Prevention published a study that looked at workers at Children's Hospital in Denver, Colorado. Of the 1000 people who received the flu vaccine before November 1, 149 still developed flu-like illness. Of the 402 people who did not get the vaccine, 68 got a flu-like illness.

a. Identify the observational units and the two variables of interest in this study. Comment on the type of variables involved. Which variable do you consider the explanatory variable and which the response variable?

b. Create a two-way table to summarize the results of this study.

c. Produce numerical and graphical summaries to describe the results of this study. Write a paragraph describing what these summaries reveal, remembering to place your statements in context and being careful how you talk about which group is being conditioned on.

d. Is there evidence that the flu vaccine helped reduce the occurrence of a flu-like illness? Explain.

Solution:

a. The observational units are people, or, more specifically, Colorado hospital workers. One variable is whether or not the person received the flu shot (categorical), and the other variable is whether or not the person got a flu-like illness (also categorical). The researchers believed that whether or not the workers received the vaccine would help explain whether or not they developed the flu-like symptoms, so the first variable is the explanatory variable and the second variable is the response.

b. The following two-way table uses Variable 1 (the explanatory variable) as the column variable and Variable 2 (the response variable) as the row variable. Your first step is to calculate how many people did not get a flu-like illness so that you can complete the row variable structure.

	Vaccine	No vaccine	Total
Flu-like illness	149	68	**= 149 + 68 = 217**
No flu-like illness	= 1000 − 149 = 851	= 402 − 68 = 334	**= 851 + 334 = 1185**
Total	**1000**	**402**	**1402**

c. If you condition on whether or not the worker received the vaccine, then you can compare the proportion of vaccine receivers who developed a flu-like illness: 149/1000 = .149 to the proportion of those without vaccine who developed a flu-like illness: 68/402 = .169.

Graphically, a segmented bar graph looks like this:

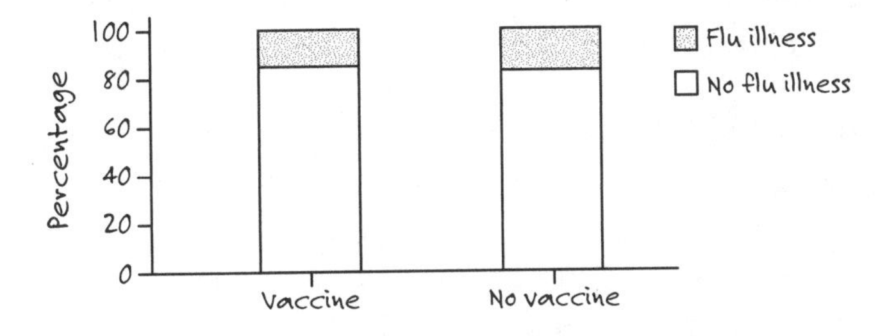

From this information, you can see that most people did not develop flu-like illnesses and that the percentages for developing flu-like illnesses were rather similar (14.9% and 16.9%) for those who received the vaccine and those who did not. The results indicate that those who did not receive the vaccine were a bit more likely to develop flu-like illnesses.

d. The proportion of individuals who developed a flu-like illness is slightly smaller for the group who received the vaccine but does not seem substantially different. Plus, it was up to the individuals themselves whether or not they obtained the vaccine. Perhaps those who did receive the vaccine were more concerned about the flu season and took other additional precautions as well. On the other hand, perhaps those with a higher risk of getting the flu were more likely to choose to receive the vaccine. Therefore, we cannot draw a cause-and-effect conclusion from this study.

KEY CONCEPTS

- When both variables are categorical, use the segmented bar graph and conditional distributions to summarize the data (rather than dotplots, boxplots, histograms, means, or five-number summaries, which are used to compare groups on a quantitative response variable). In working with segmented bar graphs, it is always preferable to scale the vertical axis in terms of percentage, have each bar total 100%, adjust for different sample sizes, and allow more immediate and more visual comparisons.
- When the conditional distributions are the same (the breakdown among the categories is the same among the groups), the variables are independent. In the vaccine example, the percentage of workers developing the illness was essentially the same regardless of whether or not they received the vaccine. These data indicate a very weak association between vaccination and development of a flu-like illness. In other words, you did not see a strong dependence between these variables.
- Even if there is a difference in the conditional distributions of the groups, as later topics will discuss further, this does not necessarily establish a cause-and-effect relationship. Another variable could be responsible. Or, when considering a third

variable, the direction of the association could even reverse, as you saw with Simpson's paradox.

- Keep in mind that *two* conditions need to be met for Simpson's paradox to occur. For example, in *WS* Activity 7-4 there was a different survival rate among fair- and poor-condition patients, *and* the proportion of fair-condition patients at each hospital differed (more went to Hospital B). In general, for Simpson's paradox a third variable needs to be related to the other two. In the hospital example, the third variable was the condition of the patients, which was differentially related to both the survival variable and the hospital variable. The following figure gives you a visual representation.

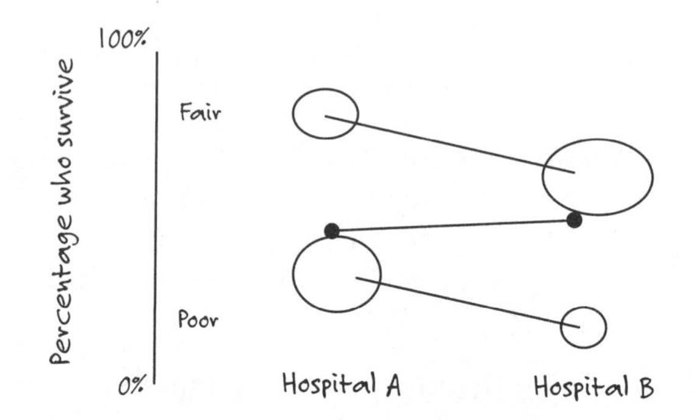

The sizes of the ovals represent the sample sizes in each group, and the center of the oval represents the overall survival rate of that group. The fair-condition patients have higher survival rates than the poor-condition patients at both hospitals. However, most of the fair-condition patients are at Hospital B, and most of Hospital B's patients are in fair condition. So when you look at the overall survival rate for Hospital B (the solid circle) it is closer to the survival rate of the fair-condition patients. But the overall survival rate of Hospital A, weighted by the large number of poor-condition patients there, will be closer to the survival rate of the poor-condition patients. In this case, the overall survival rate for Hospital A was brought below the overall survival rate for Hospital B.

- Correctly identifying response and explanatory variables is an important and often-used skill in statistical analysis.

CALCULATION HINTS

- When discussing conditional distributions, pay close attention to what you are calculating the proportion *of*. For example, "of fair-condition patients" versus "of patients at Hospital A." Practice translating your graph into a verbal description that clarifies which variable is being conditioned on. For example, are you conditioning on whether or not the person received the vaccine ("Of those receiving the vaccine, what proportion developed flu-like symptoms?") or are you conditioning on whether or not they developed flu-like symptoms ("Of those who developed flu-like symptoms, what proportion had taken the vaccine?"). You can condition on either variable (although one might read more clearly), just make sure you are consistent and that you address the question asked. Asking what proportion of 6-foot-tall men are basketball players is very different from asking what proportion of basketball players are at least 6 feet tall. If one variable is identified as the explanatory variable, we typically condition on that variable.

COMMON OVERSIGHTS

- Not using percentages on the vertical axis and therefore not stacking the segmented bar graph. Always remember to label both the horizontal and vertical axes.
- Not accumulating the conditional percentages within a category and therefore not stacking the segments to total 100% in each bar.
- Not following up graphs and calculations with a verbal description or interpretation of what they reveal.
- Not using variables along the rows and columns of a two-way table. For example, in the following table the rows are not the two possible outcomes from one variable:

	Vaccine	No vaccine
Flu-like illness	149	68
Total	1000	402

- Confusing the variables and the groups in a two-way table, as in the following table:

	Vaccine with illness	Vaccine without illness
Yes	149	851
No		

  Keep making sure that you can correctly distinguish between variables and their outcomes.

- Incorrectly calculating the conditional percentages; for example, if you were asked, "Of those with the vaccine, what proportion developed a flu-like illness?" and you responded 149/217. The value 149/217 would be the correct response to the question "Of those with flu-like illness, what proportion received the vaccine?"
- Incorrectly calculating the marginal percentages; for example, if you were asked what proportion developed a flu-like illness, you would use the results 217/1402.
- Reporting only marginal distributions and not bivariate (two variables together) distributions. For example, if the flu vaccine study had only reported that 1000 of the 1402 people had gotten a vaccine and 217 of the 1402 people had experienced a flu-like illness, that would not be enough information to investigate whether there is an association between the two variables.

WHAT WENT WRONG?

The following are examples of errors in the analysis of the example problem. In each case, explain what is wrong.

1. Suppose in *WS* Activity 7-2, part h that a student completes the bar for the 18–35 age group, as shown in the following display. What did the student do wrong?

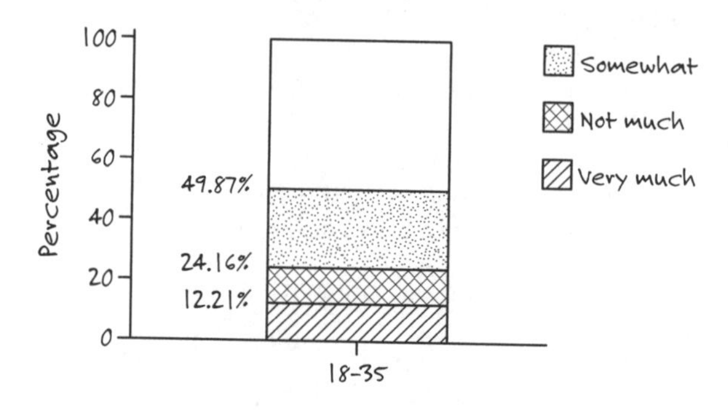

2. Suppose a student wants to study whether ice-cream preference (by flavor, chocolate, vanilla, or strawberry) is independent of gender, so he interviews 90 students and summarizes their responses in this table:

	Male	Female
Chocolate	15	30
Vanilla	12	24
Strawberry	3	6

He concludes that ice-cream preference is not independent of gender, because chocolate is preferred more than vanilla, and strawberry is preferred much less than the others. What's wrong with this analysis?

3. Suppose that Carlos is asked to investigate whether there is a relationship between whether or not he has seen a movie on the top 100 list and whether or not his lab partner Hannah has also seen it. Suppose that Carlos reports back that he has seen

37 of the movies and Hannah has seen 27. He presents the table of data in the following form. What's wrong with this table?

	Carlos	Hannah
Yes	37	27
No	63	73
Total	100	100

4. Suppose an article declares that American Airlines has a much better overall on-time arrival rate than Southwest Airlines, but when comparing the rates into Los Angeles and into Denver, Southwest has a better on-time percentage than American in each city. A student argues this is an example of Simpson's paradox and claims that the weather in Denver tends to be worse than the weather in Los Angeles. Has she sufficiently explained the cause of the paradox?

Solutions:

1. These are the correct conditional percentages, but this bar does not accumulate the percentages. The second line should be at 50.13%, the height of the first two group percentages combined, and the third line should indicate the sum of all three percentages, which would have to be 100%.
2. The "gender" and "ice-cream preference" variables are, in fact, independent in this table, because the conditional distribution of ice-cream preference is identical for the two genders: 50% prefer chocolate, 40% prefer vanilla, and 10% prefer strawberry. A segmented bar graph would show the same percentage breakdown for both bars (male and female). It is not necessary for that percentage breakdown to be 1/3, 1/3, 1/3 in order to have independence. In fact, independence versus association says nothing about how the categories within the variable compare, only how the breakdown of the variable compares between groups.
3. Carlos has confused the variables and the outcomes. If you want to know whether there is an association between Carlos's viewing history and Hannah's viewing history, then "Has Carlos seen it?" is one variable and "Has Hannah seen it?" is another. (The questions you ask about each movie aren't "Carlos or Hannah?" and

"Yes or no?" but rather "Has Carlos seen it?" and "Has Hannah seen it?") To analyze whether there is an association, you need to be able to calculate conditional proportions, for example, "Of the movies that Carlos has seen, what proportion of these movies has Hannah also seen?" Carlos's table only gives the marginal information (Hannah's outcomes and Carlos's outcomes) and does not enable you to calculate the conditional proportions to see how the variables relate. Another clue that there is a problem with the table is that there are 100 movies total, so the table total must also be 100, not 200. To properly construct the two-way table for this question, make one variable the column variable and one the row variable. However, all you can fill in based on the information from Carlos's table is the margins of the table:

	Carlos: yes	Carlos: no	Total
Hannah: yes			27
Hannah: no			73
Total	37	63	100

Once you know how many of these movies they both have seen, then you can fill in the remainder of the table and proceed with the analysis. But depending on how that number (and therefore the rest of the table) turns out, you could see an association or the two variables could be essentially independent. Below are two possibilities.

Association

	Carlos: yes	Carlos: no	Total
Hannah: yes	12	15	27
Hannah: no	25	48	73
Total	37	63	100

Essentially independent

	Carlos: yes	Carlos: no	Total
Hannah: yes	10	17	27
Hannah: no	27	46	73
Total	37	63	100

4. The student has begun a good argument but has not sufficiently explained why the direction of the association could reverse. One component could be that on-time arrival rates are higher into Los Angeles than Denver, but you would also need the two airlines to send a different number of planes to the two cities. If Southwest flew to Denver much more often than American and American flew to Los Angeles more than Denver, then Southwest's overall percentage will be pulled down by the Denver results and American's overall percentage will be pulled up by the Los Angeles results.

Topic 8:

GRAPHICAL DISPLAYS OF ASSOCIATION

When a study involves two quantitative variables, you will use the graphical displays discussed in this topic to begin your analysis. These techniques are more involved than those from the previous topic, and we encourage you to use the technology to construct these graphs for you. Doing so will free your time to focus on the analysis and interpretation of these displays.

Example: The Roller Coaster DataBase maintains a website (*www.rcdb.com*) with data on roller coasters around the world. Some of the variables recorded include whether the coaster is made of wood or steel, the structure's greatest height measured from the ground to the track level (not counting railings and flag poles), and the maximum speed achieved by the coaster. The following scatterplot displays data on the speed (in miles per hour) and height of roller coasters (in feet) in the United States for 145 steel and wooden coasters.

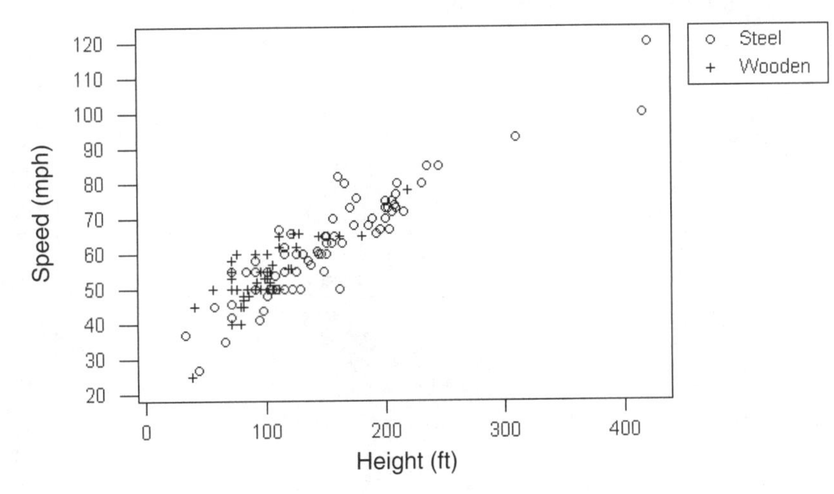

a. Identify the observational units, the variables of interest, and their types.

8

b. Describe the association between speed and height revealed in the scatterplot.

c. Describe any differences you see between the wooden and steel roller coasters.

d. Explain why the scatterplot is the appropriate graph here. Suggest some other graphs you could have constructed as well.

Solution:

a. The observational units are the coasters, and each dot in the scatterplot represents a different coaster. Three variables are examined in this graph:

- Variable 1 = speed of coaster (quantitative)
- Variable 2 = height of coaster (quantitative)
- Variable 3 = type of coaster (categorical)

b. There appears to be a rather strong, positive, linear relationship between the speed of the coasters and the height of the coasters. Two to three of the coasters are much higher than the others, but their speeds appear consistent with what we would expect based on the height/speed relationship seen among the other coasters.

c. The association between speed and height appears similar for the wooden and steel coasters. There does appear to be a larger number of steel coasters. Most of the taller and faster coasters are made of steel. The wooden coasters all occur in the lower left of the graph, indicating lower speeds and heights.

d. To explore the relationship between height and speed, two quantitative variables, a scatterplot is the appropriate graph. If you had wanted to focus more on differences based upon the type of coaster, you could have constructed boxplots of speed for the different types and boxplots of heights for the different types.

KEY CONCEPTS

- There are always at least three things to talk about with a scatterplot: direction, strength, and whether or not the pattern is linear. You should also identify any unusual observations. Remember to relate your comments to the context.
- Association between variables is another example of a statistical tendency. It is not always the case that the taller of two roller coasters is faster than the shorter one, but there is a strong tendency in that direction. In other words, there is a strong positive association between height and speed of roller coasters, but that does not mean that taller coasters are always faster than shorter ones.

CALCULATION HINTS

- When constructing a scatterplot, it is customary to place the explanatory variable on the horizontal axis and the response variable on the vertical axis. In some situations, as with the coasters, this distinction may not be the primary focus of the research study.
- When graphing "*A* versus *B*," the convention is to place variable *A* on the vertical axis and variable *B* on the horizontal axis.

COMMON OVERSIGHTS

- Overstating your conclusions; for example, "The steel coasters are taller." The data show a pattern that the steel coasters *tend* to be taller, but there are some short ones as well.
- Talking inappropriately in terms of time when interpreting the scatterplot; for example, "The coasters are getting faster," or "The speed of the coasters is going up." These are sensible statements considering the impact of technology on roller coasters over time, but the scatterplot from the example doesn't support these conclusions. If the horizontal axis had been "year built" with the same positive association, then these conclusions would have been valid. Although it is common to say that a graph like this is "going up," in a statistics context you must speak in terms of the relationship between the variables displayed.
- Talking inappropriately in terms of cause and effect; for example, "As the coasters get taller, they get faster." This also makes some intuitive sense but is not quite what this scatterplot reveals. You don't know why the coasters are faster or when they

started to get faster, you only know that taller coasters tend to also be faster than shorter coasters.

- Constructing a scatterplot even when the two variables are not recorded on the same observational units or are not both quantitative.

WHAT WENT WRONG?

The following are examples of errors in the analysis of the example problem. In each case, explain what is wrong.

Suppose that an instructor teaches a class with 10 men and 10 women, and he records these scores on the first exam (listing scores by student name in alphabetical order within each gender):

Men:	85	72	84	98	66	59	74	80	90	81
Women:	95	72	83	77	93	88	77	83	65	99

He constructs the scatterplot on the left below and concludes that there is no association between men's scores and women's scores. Then he sorts the scores for each gender, produces the scatterplot below right, and changes his mind to conclude that there is a strong, positive association between men's scores and women's scores. What is wrong with these analyses and conclusions?

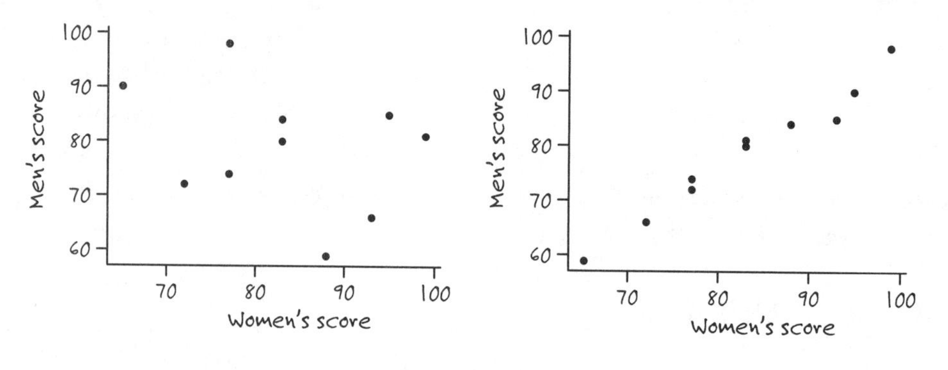

Solution:

Neither of these graphs makes any sense. In a scatterplot, each dot must represent one observational unit, with two quantitative variables measured on that observational unit. In this situation the observational units should be the students, but only one quantitative variable (exam score) has been measured on each student. The second variable here is gender, which is categorical. A sensible graph would be to create boxplots of the exam scores, classified by gender.

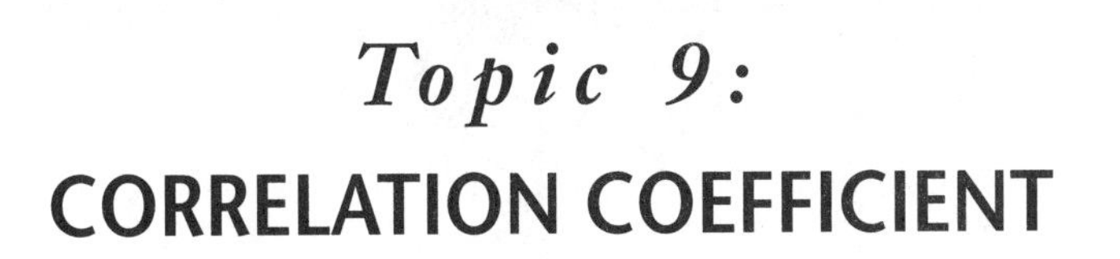

Topic 9: CORRELATION COEFFICIENT

The previous topic examined the appropriate graphical display to explore the relationship between two quantitative variables. As always, the next step is a numerical summary: the correlation coefficient. Again, the focus is on the direction, strength, and linearity of the relationship.

Example: Recall the roller coaster speeds and heights from the previous *TK* topic:

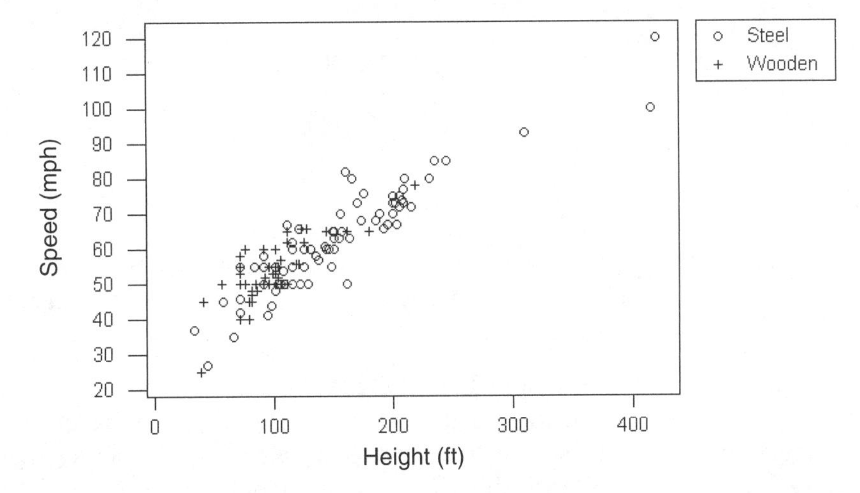

a. Guess the value of the correlation coefficient between speed and height for these roller coasters. What are the measurement units on this calculation?

b. If you were to calculate the correlation coefficient separately for the wooden and steel roller coasters, how do you think they would compare? Explain.

c. If you remove the two tallest roller coasters from the data set, how do you think the correlation coefficient from part a would change?

d. Suppose that the two tallest coasters had maximum speeds of only 70 mph. How would you expect the correlation coefficient to change, and what would happen to its usefulness?

Solution:

a. With such a strong positive linear relationship, you would expect the value of the correlation coefficient r to be fairly close to 1. The correlation turns out to be $r = .895$ (and is a unitless value).

b. The relationship between speed and height for the wooden coasters appears slightly weaker than that for the steel coasters, though both are quite strong. The relationship for the steel coasters extends over a broader range, giving further evidence to the relationship. This phenomenon will also create a larger value for the correlation coefficient. For the steel coasters $r = .910$, and for the wooden coasters $r = .773$.

c. Because the two coasters in question are still roughly in line with the overall linear pattern, removing them would probably not greatly change the value of the correlation coefficient for the remaining data. If anything, you might expect that removing points that are in line could lower the value of r. It turns out that with these two coasters removed $r = .897$, essentially the same as in part a.

d. In this situation, the correlation coefficient would be quite a bit smaller (it becomes $r = .777$). Moving those two points would create an overall relationship that is much weaker, and would raise serious concerns about calculating the correlation coefficient because the relationship would no longer be linear.

KEY CONCEPTS

- The correlation coefficient indicates the strength and direction of the linear relationship between two quantitative variables. This calculation should not be applied if the relationship is not linear, so check your scatterplot first! A key feature of the correlation coefficient is that it will not change in value if you rescale either of the variables (e.g, feet to meters). In addition, r is unaffected by which variable is considered the explanatory and which the response.
- One caution with the correlation coefficient is that it is highly sensitive to extreme observations (i.e., it is not resistant) because it is based on x and s, which are not resistant. If there are a few unusual observations, the value of r could be rather misleading. In this case, it is often useful to report r both with and without the unusual observation(s).
- Even if two variables have a very strong association, this does not enable you to conclude that one variable is therefore causing the other.

CALCULATION HINTS

- Always check the reasonableness of a calculated value. For example, r can never be larger than 1 or smaller than -1. *Measurement units* are not used with r.
- Remember not to try to calculate r with data that are not quantitative. You should also be very cautious in using r with nonlinear data. In addition, consider the possible effect of different clusters on how well the correlation coefficient summarizes the overall relationship.

COMMON OVERSIGHTS

- Mistakenly assuming that a relationship has to be strong to be linear, and vice versa. Variables can have very strong nonlinear relationships and very weak linear relationships. You saw an example of the first case in *WS* Activity 9-1, part g. There it would not be appropriate to use r as a measure of the overall strength of this nonlinear relationship. In the latter case, it is appropriate to calculate r, and we would expect r to be close to zero, as shown below. (The lines have been added to the graph to help you visualize the overall linear pattern.)

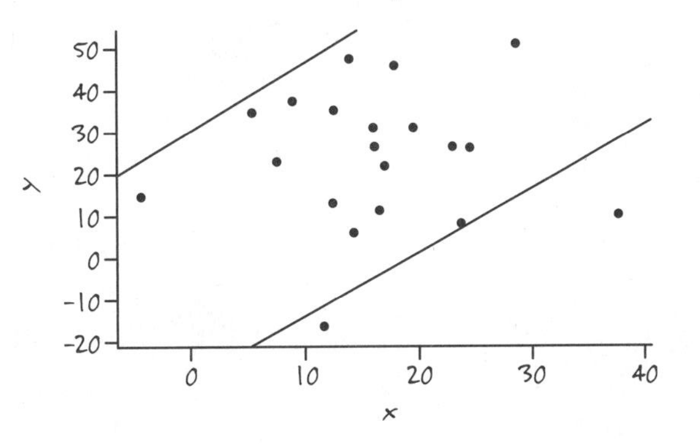

Topic 10:
LEAST SQUARES REGRESSION I

After summarizing the relationship between quantitative variables graphically and numerically, it is often appropriate to model that relationship. If you can suggest an equation for a line that reasonably summarizes the overall pattern, then you can use this line to make predictions of the response (y) variable for any value of the explanatory (x) variable. Again, it's best to let technology do much of the "heavy lifting" so that you can focus on the concepts and the interpretations.

Example: Recall the roller coaster speeds (mph) and heights (feet), classified by type from the previous *TK* topics:

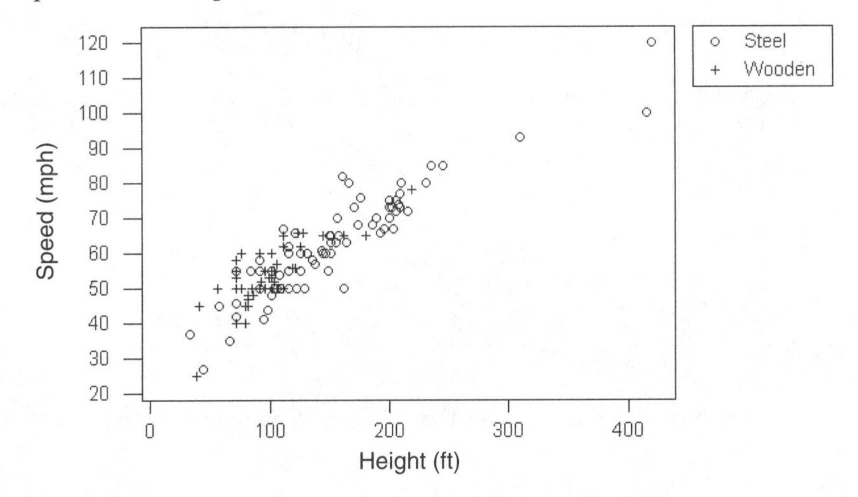

a. Take a guess as to the equation for the least squares line for the full data set. (Write this out in full equation form.)

The least squares equation is $\widehat{speed} = 34.1 + .19\ height$.

b. Provide an interpretation for both the slope coefficient and the intercept coefficient.

c. Use the least squares line to predict the maximum speed for a coaster with height 350 feet.

d. What would you conclude about the speed of a coaster with height 600 feet?

e. Circle one or two points on the graph that you believe will have large (in absolute value) residuals.

f. The r^2 value for the full data set is 80.1%. How do you interpret this value?

g. Suggest how separate least squares lines for the wooden coasters and for the steel coasters might differ.

Solution:

a. First, trace a line across the graph that you believe summarizes the overall relationship:

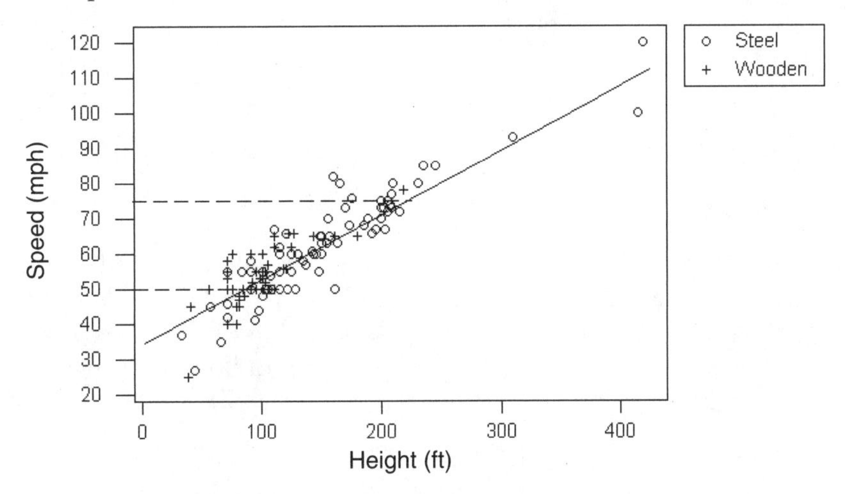

You know that the slope needs to be positive and that the y-intercept is found by determining the height of this line when the roller coaster height is 0 feet. From the line sketched in the plot, this appears to be around 35 mph. If you eyeball the speeds indicated by this line for coaster heights of 100 and 200 feet, the difference appears to be around 75 − 50 = 25 mph. This estimates the rate of change to be 25/(200 − 100) = .25 mph/ft. Putting these together, the least squares line is $\widehat{speed} = 35 + .25\ height$. [*Note:* Technology reveals that the actual least squares line is $\widehat{speed} = 34.1 + .19\ height$, quite close to the eyeball estimate.]

b. The slope indicates that if a coaster is 1 foot higher than another, you will predict the maximum speed to be .19 mph faster for the taller coaster. Thus, if one coaster is 100 feet higher than another, you predict its maximum speed to be 19 mph faster. Technically, the intercept (34.1 mph) predicts the maximum speed for a coaster of height 0 feet. Clearly this does not make much sense in this context, but that's what the intercept means.

c. The predicted speed for a coaster of height 350 feet is $\widehat{speed} = 34.1 + .19(350) = 100.6$ mph.

d. It would be very risky to make a prediction for a 600-foot coaster using this equation because the coefficients were determined based on data of coasters up to 400 feet tall. You don't have any information about how the relationship might change for coasters taller than 400 feet.

e. The following graph shows the least squares line. The points circled are farthest from the line in the vertical direction, representing the largest (in absolute value) residuals.

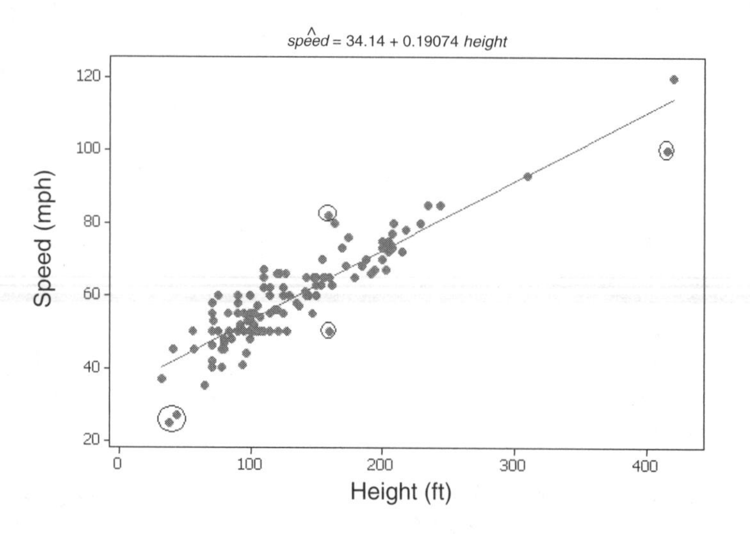

f. An r^3 of 80.1% means that 80.1% of the variability in roller coaster speeds is explained by the regression line on roller coaster height. As can be seen in the preceding graph, there is not much deviation of the observations away from the line. This line is a pretty good predictor of how fast a coaster will go given its height.

g. The following plot is an attempt to show one line for the wooden coasters and one for the steel coasters.

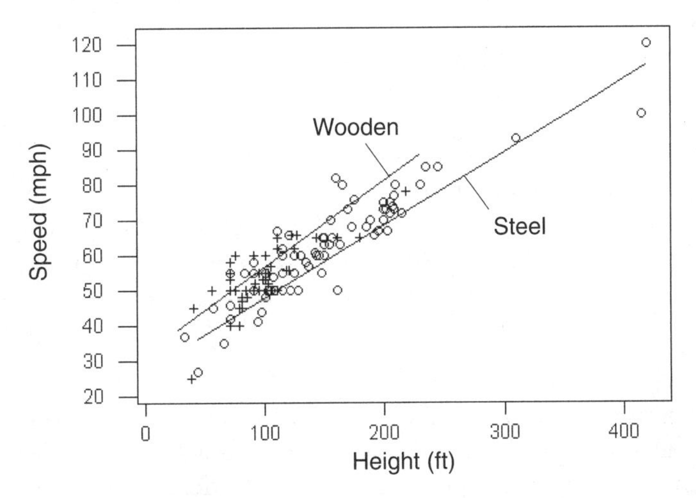

The slope for the wooden coasters appears slightly larger than the slope for the steel coasters because its line is steeper. The discussion in *TK* Topic 9 showed that the relationship between height and speed is for wooden coasters, but that does not imply that the slope will be smaller, as you see here.

Technology reveals that the two equations are:

Wooden: $\widehat{speed} = 34.3 + .204\ height$

Steel: $\widehat{speed} = 32.3 + .198\ height.$

The difference in the slopes is not at all large.

KEY CONCEPTS

- A least squares regression line summarizes the overall linear relationship between two quantitative variables. It can be used to make predictions about the response (y) variable for different values of the explanatory (x) variable (as long as the x-values we want to predict for are not outside the range of the x-values in the data set). [*Note:* We use the terms "regression line" and "least squares line" interchangeably.]
- The least squares regression line is calculated by minimizing the sum of the squared residuals of the data points from the line. Because we are trying to minimize the size of the residuals (in the vertical direction), the equation can be sensitive to the effects of extreme values, especially if they are extreme in the x direction.
- It matters a lot to the regression line which variable is treated as the explanatory and which as the response. Make sure the variable you are trying to predict is treated as the y-variable.
- The slope of the line tells us the *average* (or *predicted*) change in y if the x-variable increases by 1 unit. The intercept tells us the predicted value of y when $x = 0$.

- The r^2 value tells you the percentage of variability in the response variable that is explained by the regression on the explanatory variable. It gives a measure of how much better your predictions will be if you take the explanatory variable into account versus only using the mean of the y-values to make a prediction.

CALCULATION HINTS

- Make sure the numbers you obtain make sense in the context. For example, make sure that the slope is positive if the scatterplot reveals a positive association; and make sure that predictions look reasonable, based both on common sense (a predicted coaster speed of 500 mph would not be reasonable) and on the scatterplot (a predicted speed of 40 mph for a 200-foot coaster would look out of place in the scatterplot).
- Be very careful with rounding. When using the equation $b = rs_y/s_x$ for the slope and $a = \bar{y} - b\bar{x}$ for the intercept, calculate to as many decimal places as possible and only round off when reporting the equation itself. In particular, when calculating the intercept a, use as many decimal places as possible in the slope b, not the rounded-off version.
- If you are given r^2 and asked to calculate the correlation coefficient r, remember that r could be either positive or negative. You can determine which by referring back to the scatterplot and seeing whether the association revealed there is positive or negative.

COMMON OVERSIGHTS

- Switching the roles of the two variables. Be sure to start by identifying which variable is the response (y, the one being predicted) and which is the explanatory (x, the one being used to make the prediction). Then be sure to graph the variables, calculate the least squares line, and make interpretations consistent with your choice.
- Not writing out the least squares equation in terms of the variables involved (e.g., saying "y" and "x" instead of "speed" and "height") and not putting a "hat" over the response variable.
- Mixing up fitted value, residual, and the observed value. It's important to differentiate between the actual data value (y) and the value predicted by the regression line for that observational unit ($\hat{y}$). The former was actually observed, the latter is just a prediction. The residual is always the difference between what was observed and what was predicted. Keep in mind that points above the line have positive residuals, so you subtract; $y - \hat{y}$. Another way to think of this is with the equation *actual* = *fit* + *residual*. In other words, the fitted value is based on the overall model, but the actual values will differ from that model by a (hopefully) small residual (or error).
- Being sloppy in the interpretation of r^2. The definition of r^2 is difficult to grasp, and it is very easy to leave out a word or two, making the interpretation incorrect. It is tempting to try to memorize the generic definition, but it is important to be able to state the interpretation in the context of the data. We encourage you to practice interpreting r^2 and then have someone else review your response in detail.

- Not thinking about the predicted response or the average response when interpreting the slope.
- Forgetting about context and units in your interpretations; for example, saying that the predicted speed of a 200-foot coaster is 72. Say instead that the predicted speed is 72 miles per hour.

WHAT WENT WRONG?

The following are examples of errors in the analysis of the example problem. In each case, explain what is wrong.

Consider the following scatterplot of the time (in minutes) and distance (in miles) for hikes described in the book *Day Hikes in San Luis Obispo County*:

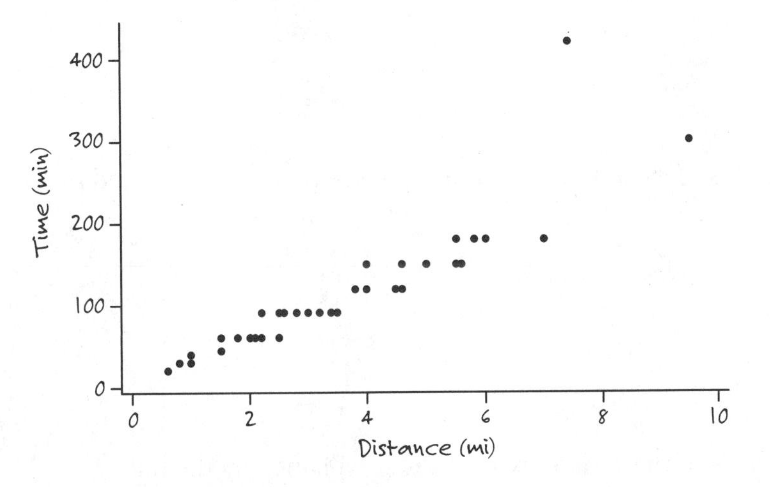

1. Suppose that a student calculates the least squares line to be $\widehat{time} = .564 + 0.27\ distance$. Explain why this equation must be wrong, and make a guess as to how the student mistakenly calculated this equation.

The correct least squares line for predicting time from distance is $\widehat{time} = -1.27 + 31.5\ distance$. The value of r^2 is .838. Identify at least one problem with each of the interpretations and conclusions in parts 2–8.

2. The slope shows that for each additional minute, we predict the hike is 31.5 miles longer.

3. The slope shows that a hiker's time increases by 31.5 minutes for each additional mile.

4. The predicted time for a 5-mile hike is about 156.

5. About 84% of the hikes have times that are correctly predicted by the line.

6. About 84% of the variability in hikes is explained by the line.

7. The 9.5-mile hike has the largest residual.

8. The 7.4-mile hike has the largest fitted value.

Solutions:

1. This slope is not reasonable because it means that the predicted time of the hike increases by only .027 minutes (less than two seconds!) for each additional mile of distance. Another way to tell that this equation is mistaken is that the predicted time for a 5-mile hike would be $\widehat{time}$ = .564 + .027(5) = .699 minutes, which would be the fastest 5-mile hike ever recorded! The student made the error of switching the roles of the variables in this calculation.

2. The interpretation of the slope is incorrect. The student has reversed the role of the x- and y-variables. When interpreting the slope, you discuss how a unit increase in the *explanatory* variable changes the predicted value of the response variable.

3. This interpretation of the slope is too definitive; it makes it sound as if there is no variability present. A better interpretation would be that the predicted time of a hike increases by 31.5 minutes for each additional mile, or that the hike time increases on average by 31.5 minutes for each additional mile. Although the y-value of the corresponding position on the line changes by the exact same amount each time we move 1 unit in x, this is not exactly true for the data values themselves. The interpretation of the slope applies to the line and not to individual hikers.

4. The calculation is correct here, but what's missing are the units: the predicted time for a 5-mile hike is 156 *minutes.*

5. This is a completely incorrect interpretation of r^2. It may well be that none of these hikes have a time that is perfectly predicted by the line. That is not what r^2 means.

6. This interpretation of r^2 comes closer but is still lacking because it does not indicate what variable about the hikes is being explained. A good interpretation is that about 84% of the variability in the *times* of the hikes is explained by the least squares line based on the hikes' distances.

7. The 9.5-mile hike has the largest distance (x) value, but the line comes fairly close to this point, so it does not have a large residual. The 7.4-mile hike that takes more than 400 minutes has the point falling farthest from the line and therefore the largest residual.

8. The 7.4-mile hike has the largest *actual* time, but not the largest fitted value. Remember that a fitted value is a predicted value, so the positive association in the data means that the largest fitted value belongs to the hike with the largest distance. In this case, the 9.5-mile hike has the largest fitted value.

FURTHER EXPLORATION

Open the "Behavior of Regression Line" applet at *www.rossmanchance.com/applets/LRApplet.html*. You will see a scatterplot and least squares line for some generic data.

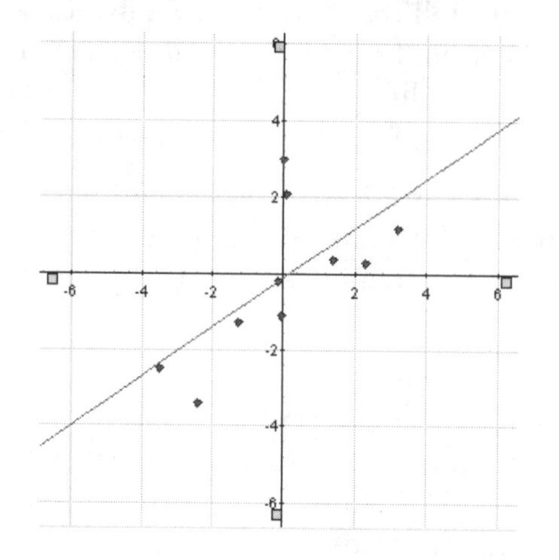

a. Report the least squares regression line and the correlation coefficient.

b. Click on the point at (3.27,1.24). Hold the mouse button down and drag the point up vertically. What happens to the least squares line? What happens to the correlation coefficient?

c. Drag the point downward; is it possible to make the slope of the least squares line negative?

d. Return the first point to about where it started. Select the point that is close to the origin (−0.13,−0.17). Drag this point up and down (trying to keep the x-value about the same). Are the changes in the line and correlation coefficient as dramatic as in part b?

e. Return the second point to its starting position, and click on the first point (it should turn yellow). Press the **BACKSPACE** key to delete the first point. Record the change in the least squares equation and the correlation coefficient. Now repeat for the second point.

f. Summarize what this applet has shown you about potentially influential observations.

Solution:

a–f. An influential observation is an observation whose removal from the data set changes the least squares line appreciably. This exercise should have reminded you that points that are extreme in the x direction are more likely to be influential than points near the mean $\bar{x}$. In fact, one observation extreme in x can single-handedly determine whether the slope is positive or negative.

Topic 11:
LEAST SQUARES REGRESSION II

This topic continued your study of regression by further exploring influential observations. You also learned about using transformations to make bivariate data more linear.

Example: Recall the scatterplot and least squares line for predicting the roller coaster speeds from their heights.

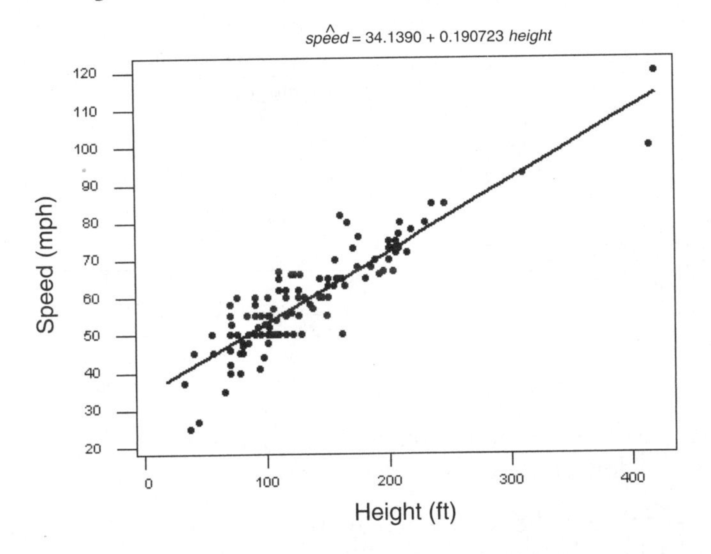

a. Suppose the two coasters with the greatest heights had maximum speeds of only 70 mph. How would that change the linear regression equation?

b. Which coasters do you think are the most influential? Explain.

c. The following plot displays the residuals versus the height of the coasters. Summarize the information the graph reveals.

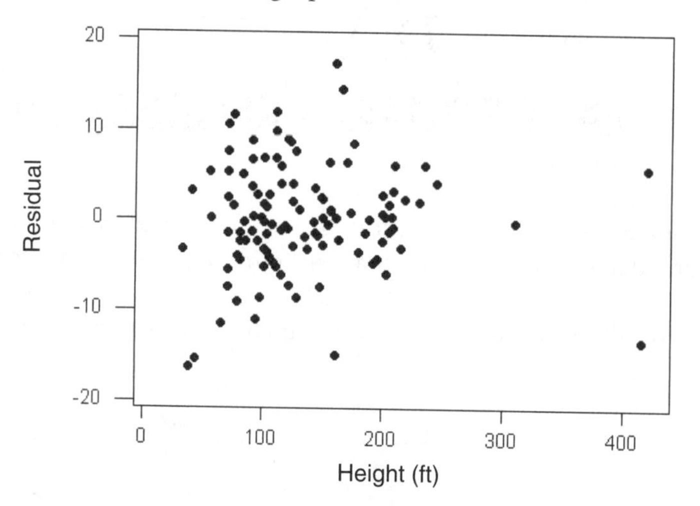

Solution:

a. If those two observations are made that much lower, they will "drag" the least squares line down to have a much smaller slope.

b. The two or three roller coasters that are much taller than the rest will be most influential because their heights (the x-variable) are so far from the other heights.

c. The residuals look rather scattered, and there is no evidence of curvature. The unusual roller coasters, those with larger residuals or extreme heights, stand out even more in this graph. There may be some evidence of smaller residuals with taller roller coasters, but this is hard to determine because of the small number of coasters with large heights.

COMMON OVERSIGHTS

- Mixing up or equating observations that have large residuals and potentially influential observations. These are two different phenomena. Residuals deal with an observation's distance from the regression line; influence concerns how much the equation for the line would change if that value was removed. Points have a higher potential for influence when their x-value is extreme. Truly influential observations may even have a small residual, because they pull the line toward them.

- Assuming all residual plots look fine. It takes some practice to know what a problematic plot looks like, but don't automatically assume everything is ok. The main thing to worry about now is evidence of curvature, because that indicates that the regression line may not be the most appropriate model for your data.
- Neglecting to transform a variable when making predictions based on a model that used a transformation. For example, in *WS* Activity 11-3 you should have arrived at the model $\widehat{life} = 80.6 - 13.3 \log(people)$. Then, to predict the life expectancy of a country with 100 people per television, be sure to take the log of 100 before proceeding.

WHAT WENT WRONG?

The following are examples of errors in the analysis of the example problem. In each case, explain what is wrong.

Runners are concerned about their form when racing. One measure of form is the stride rate (number of steps taken per second). A runner is inefficient when the rate is either too high or too low. As speed increases, so should stride rate. In a 1977 study, 21 of the best American female runners were timed, and their speed (feet per second) and average stride rate were recorded. The following scatterplot shows the relationship between stride rate and speed for these runners. A student concludes that it would be perfectly appropriate to model this relationship with a line. Do you agree?

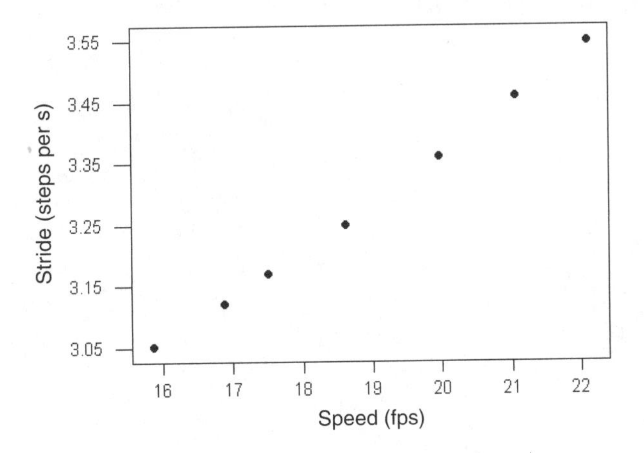

Solution:

Although the scatterplot appears very linear, it's always a good idea to check the residual plot before proceeding.

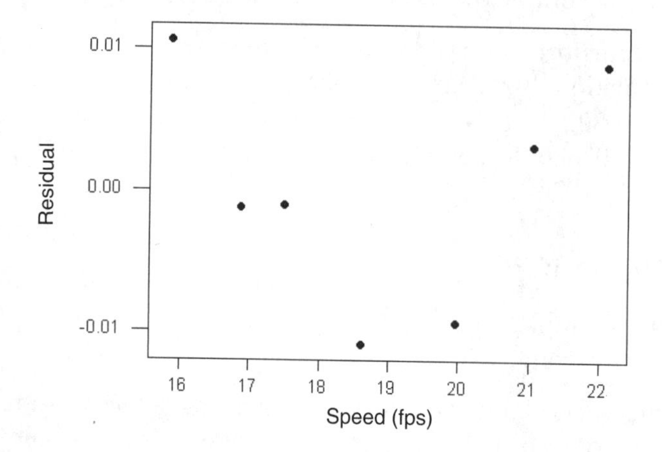

This plot indicates there is some curvature to the relationship that would be missed if we automatically fit a line to these data.

Unit III:

Collecting Data

Topic 12:
SAMPLING

In the previous topics, you developed tools for describing and analyzing data. Underlying this analysis is the assumption that you have "good data." In *WS* Topic 12 (and 13), you learned how to properly obtain data to address the question of interest. This is a crucial first step in any investigation, and you should always consider how the data were collected when you draw your final conclusions. In this topic, you also began to learn about making inferences beyond the data you have in hand. There are two types of inferences we will be concerned with:

12

1. Inferences from a sample to a larger population. If the observational units you observe/measure are a subset of observations from a larger group, what statements can you make about the larger group based on the data in hand? *WS* Topic 12 discusses appropriate techniques for selecting observations from the larger population so that such inferences are appropriate.
2. Cause-and-effect comparisons between groups. Previous topics discussed using caution before thinking that an association is causal. In *WS* Topic 13 you will learn how to design a study so a (stronger) conclusion of causation is warranted.

Example 1: Suppose you want to find out about sports fan opinions of Pete Rose. The Gallup Organization conducted such a poll on January 9–11, 2004. They asked,

> *Next, we'd like to get your overall opinion of some people in the news. As I read each name, please say if you have a favorable or unfavorable opinion of these people—or if you have never heard of them. How about . . . Pete Rose?*

One proposal is to ask the first 1000 people leaving a Los Angeles Lakers' basketball game.

a. Identify the observational units and the variable of interest. Is this a quantitative or a categorical variable?

b. Define the population and the sample being considered.

c. Is this a reasonable sampling plan? Explain.

d. An alternative proposal is to select 1000 people from each National Basketball Association game in the United States that weekend. Now the sample size is about 20,000 people. Will this approach address the problems you identified in part c?

e. Suppose you had a list of all subscribers to *Sports Illustrated*. Explain how you could use this list to select a random sample. Describe the population that would be represented by this random sample.

f. When Gallup poses a question to their sample, they often rotate the order of the responses to the question, or the order of the people that they ask about (e.g., Pete Rose will not always be the first person asked about). Why do you think they do this (e.g., favorable and unfavorable)?

g. In Gallup's poll, they found that 49% of the respondents had an unfavorable opinion of Pete Rose. Define the parameter and the statistic for this study.

h. If Gallup had selected another sample of 1000 people and asked them, which of your answers to part g would change?

Example 2: The following excerpt describes a study on the relationship between eating sweets and how long people live. Read this excerpt and answer the questions that follow.

Indulge, Chocolate Lovers – You'll Live Longer, Researchers Say

By Emma Ross
THE ASSOCIATED PRESS

LONDON — Scientists already have suggested that eating chocolate may make you happy. Now they say men who indulge in chocolate may live longer.

A study of 7,841 Harvard male graduates found that chocolate and candy eaters, regardless of how voracious their appetite for goodies, live almost a year longer than those who abstain.

The researchers from Harvard University's School of Public Health, whose study was published in this week's issue of the British Medical Journal, said they don't know why this is. They speculate, however, that antioxidants present in chocolate may have a health benefit.

The scientists stress their findings are preliminary, and other experts caution that the research does not prove the results can be attributed to the antioxidants.

In the study, those who ate a "moderate" amount of sweets, allowing themselves only one to three candy bars a month, fared the best, having a 36 percent lower risk of death compared with non-candy eaters.

Although they fared worse than the moderates, the more ardent confectionery eaters, classified as those treating themselves to three or more sweets a week, still lived longer than those who had banished candy from their lives, with a 16 percent decreased risk of death.

Scientists previously have found that chocolate contains phenols, antioxidant chemicals also present in wine. Antioxidants prevent fat-like substances in the blood from oxidizing and clogging the arteries.

"It's raising a hypothesis that, if true, would bring cheer to those who like chocolate," said Dr. I-Min Lee, an epidemiology professor at Harvard who led the research.

The study, in which men were questioned in 1988 about their candy habits over the past year, did not count chocolate sauce, cake, or other chocolate consumed in desserts.

The men, who started the study at an average age of 65, were followed for five years, by which time 514 had died.

The results took into account factors including weight, smoking and other lifestyle habits, but Lee still warned that other health effects could have played a role.

Andrew Waterhouse, a wine chemist at the University of California at Davis who has researched chocolate's antioxidant properties, said the Harvard hypothesis that the decrease in death risk could be linked to chocolate is "plausible."

He said he was surprised, however, by the finding that eating only three chocolate bars a month was associated with a lower death risk and doubted that was due to phenol.

"That frequency is so low it was surprising there would be any benefit from an antioxidant effect, if that's the full explanation," he said. "Oranges are high in antioxidants and you wouldn't expect much benefit from eating three oranges a month."

Dr. Catherine Rice-Evans, a professor of biochemistry at Guy's Hospital in London, also doubted that the phenols in chocolate could explain the results.

She said she also was suspicious that the study did not find that the more chocolate the men ate, the lower their risk of death.

"That observation suggests to me that it's nothing to do with the chocolate," she said, adding that studies controlling the diet are necessary to discover if it is a real effect of chocolate.

Source: *The Record,* Bergan County, New Jersey, December 18, 1998.

a. Identify the population and the sample discussed in this article.

b. Identify the two primary variable(s) of interest and the parameter(s) of interest.

c. Explain some positives and negatives about how the researchers selected the individuals to participate in their study. In particular, discuss a possible bias in this sampling method.

Solutions:

1. **a.** The observational units are sports fans. The variable of interest is whether their opinion is favorable (categorical).

   b. The population is all sports fans. The sample consists of those questioned by Gallup.

   c. No, the people leaving a Lakers' game will not be representative of all sports fans. First and foremost, this sample includes primarily individuals from California, who do not represent the entire nation. Furthermore, we have only asked people at a basketball game, so fans of other sports are not necessarily represented. People who left the game early may also differ in their opinions about sports than those who stayed for the entire game.

   d. Although this does increase the sample size and has more representation from other states, it does not do much better at representing all sports fans across the nation. It still only includes people attending basketball games and those who chose to attend that weekend. There are many other types of sports fans who will not have had an opportunity to participate in this poll. Much more important than the sample size is how the people are selected for the poll.

e. If you treat the *Sports Illustrated* subscribers list as the sampling frame, then you could number every name on the list and use the random number table to randomly select 1000 to be in the sample. You would then need to contact those people to ask their opinion. Ideally, you would perform repeated call backs of any individuals who were not home the first time. Still, you would probably only consider this sample representative of *Sports Illustrated* subscribers and not of *all* sports fans. This approach does eliminate the selection bias of those who chose to attend a basketball game. As the Gallup website states, the first thing they do when conducting a national poll is to select a place where all or most Americans are likely to be found. This will not be a mall, a hotel, or a sporting event. Instead, they begin all of their national surveys at people's homes.

f. How a question is asked can be another source of bias. If you try to equalize any potential influences from question wording or even question order, this should provide a more representative response to the question of interest across the sample. As the Gallup website states, it is important to always include the exact wording of the question(s) used.

g. If 49% of the sample responded this way, this 49% is a statistic. The corresponding parameter is the percentage of the entire population who would respond this way. You do not know the value of this parameter.

h. The definitions of the sample and the parameter would stay the same, but the value of the statistic would probably differ because there is variability in the sample results from sample to sample. However, the value of the parameter is constant (albeit unknown to you) as you take different samples.

2. a. The population claimed by the newspaper article is "men." The sample is 7841 Harvard male graduates.

b. The primary variable of interest is how long each person lives or whether or not the person has died. The corresponding parameter could then be the average lifetime of men or the proportion of men who die within 5 years. In this case, you would actually define two parameters, the average lifetime of men who eat chocolate and the average lifetime of men who do not eat chocolate. You are interested in estimating the difference between these parameters. Alternatively, you could define one parameter to be the proportion of all male chocolate eaters who die within 5 years and a second to be the proportion of all men who do not eat chocolate who die within 5 years.

c. Negatives: It is very debatable whether the results for Harvard male graduates will be representative of all males. There could be something different about these individuals and their life expectancy compared to the general male population. For example, Harvard male graduates are probably wealthier and better able to maintain high dietary standards and medical care than other men. This difference would lead to a greater average life expectancy when the sample is collected this way than if this were a true random sample of all males.

Positives: If the real issue here is the comparison of those who eat sweets and those who don't, you will be able to make a better comparison of this variable when the individuals in the study are similar to begin with. It seems unlikely that the effect of sweets on health would be different for Harvard graduates than for other men, but it is certainly possible that the health effects of chocolate on women could be different than on men, so the population here does not include women.

KEY CONCEPTS

- The terms sample, population, parameter, statistic, and sampling frame are fundamental to statistical analysis.
- You must take care in how you select a sample from a population if you hope to generalize information from the sample to describe the larger population. For example, in *WS* Activity 12-3 if you selected the first senators you thought of (senators by memory), you would tend to overestimate the years of service in the entire senate.
- A simple random sample gives every member of the population an equal chance to be in the sample. By selecting simple random samples from the population, the distribution of the statistics will center around the population parameter, eliminating bias from the sampling method. [*Note:* We are talking about the bias in the *method*, not in an individual sample.] For example, if you took several random samples of senators from Activity 12-3, you would get different results for the average years of service from sample to sample, but the center of this distribution would be at the population mean.
- Don't forget about other "nonsampling" sources of bias, such as wording of the question, appearance of the interviewer, and so forth.
- Increasing the sample size reduces the amount of sampling variability but will not reduce sampling bias. For example, the sample means from different samples will tend to cluster close together when the sample size is large. If your sampling method is biased, then increasing the sample size will not help. The sample means may cluster closer together but will still cluster around the wrong value. You will have a more exact answer, but to the wrong question!
- The population size is not relevant in determining whether there is sample bias and, in most cases, is not relevant in determining the precision of the sample statistics.

CALCULATION HINTS

- Be careful in describing your set-up when using the random number table. Make sure each number has the same chance of occurring. For example, if you numbered the list 1–100, you need to use three-digit numbers in your search.

COMMON OVERSIGHTS

- Talking about selecting the sample "at random" and not relying on the statistical definition of the random sampling process, that is, numbering every individual in the population and using a random number table or other random mechanism to take the sample. It is easy to throw the word "random" around, but be sure that when you use the phrase, you really mean it. Don't assume that everything in the data collection process has been done reasonably and be *very* cautious in generalizing your results to the larger population unless the sample was randomly selected.

- Confusing statements of the sample, the population, and the variable or the parameter. For example, in the sweets study, saying the sample is "those who eat chocolate." You've jumped ahead a bit; the observational unit is people, and the variable is whether they eat chocolate. Though the study looked at *how many* men died, the "question" asked of each individual is *whether or not* they died.
- Not stating the sample as a subset of the population. Both are a collection of the same observational units. For example, it is not quite consistent to say the population is all U.S. households and the sample is people who responded to the phone survey, because this response refers to households in one case and people in the other. You should say households for both or people for both. Households would probably be better in this case since the data collected are typically at the household level in phone surveys.
- Not arguing for a particular direction when discussing bias. Don't just say, "There is bias," but discuss whether you think the method used will systematically lead to an overestimate or an underestimate. Don't just say, "The results will be off," or "The results could be higher or lower." Pick one direction and argue for that direction. Remember, bias is a systematic tendency to err in a particular direction, not just "to err."
- Confusing bias and precision, and believing that taking a larger sample reduces bias. If the sampling method is biased, then taking a larger sample will not reduce the bias. In fact, taking a larger sample will exacerbate the problem by producing a more *precise* estimate that is still biased. The infamous *Literary Digest* poll of 1936 is a prime example of this phenomenon: That poll produced a very misleading, biased result despite its enormous sample size.
- Not thinking of the sampling frame as a physical list. For example, saying the sampling frame is "all Harvard graduates" instead of "a list of all alumni from the university."

WHAT WENT WRONG?

The following are examples of errors in the analysis of the example problem. In each case, explain what is wrong.

1. A random sample of 100 students at a local high school is asked how many people are in their household. Will this be a good estimate of the true average size of households in your state?

   a. Yes, the sample was selected at random.

12

b. No, the sample is too small a percentage of the state population.

c. No, if the family has school-age kids it could be larger or smaller than the other families in the state. You have missed those families without kids and families with younger or older children.

2. Suppose that you want to estimate the proportion of Americans over the age of 12 who have heard of the Gallup Organization, so you take a sample of 50 students in an introductory statistics course at a local high school. Consider the following student reaction: "This sample is biased. It's not random because not all American adults study statistics. Also a sample of size 50 couldn't possibly represent all Americans over the age of 12." Identify some flaws and omissions in this response.

Solutions:

1. **a.** The sample was selected at random but only from the population at the school, not from the households across the state. This is especially problematic if certain types of families have children at this school (e.g., families at a specific socioeconomic level) and that factor is related to how large the family tends to be.

   b. The population size is not relevant for determining whether the sample is representative.

   c. This is a very good answer, but it argues in both directions. We encourage you to pick one direction and support that direction only. If you really had overestimates and underestimates in repeated samples, wouldn't they balance out so that the results would still be centered in the right place? You should consider all possible sources of bias but focus on what you think would be the worst cause of bias or several possibilities that together could cause a significant error in the same direction.

2. First, the problem with this study is not simply that not all Americans over age 12 study statistics. A more complete description of the problem is that it seems reasonable to expect that people studying statistics would be more likely to have heard of the Gallup Organization than the general public. Thus, this sampling procedure probably overestimates the value of the parameter (the proportion of American adults who have heard of the Gallup Organization). A more subtle problem with this student's response is that it refers to the *sample* as being biased, but bias is a characteristic of the sampling *method*, not of one particular sample. The sample size and population size are also irrelevant when considering the issue of bias. If this had been a true random sample of all Americans over the age of 12, then even small samples will be "representative," they just will not be very precise.

FURTHER EXPLORATION

In *WS* Activity 12-3 you considered taking samples from the 1999 U.S. Senate.

a. What was the expected direction of the bias when we asked you to think of five senators you had heard of?

b. The "Sampling 2002 Senators" applet at *www.rossmanchance.com/applets/senators/samplesenators.html* contains information about the 2002 U.S. Senate.

12

Simulating Senators Samples

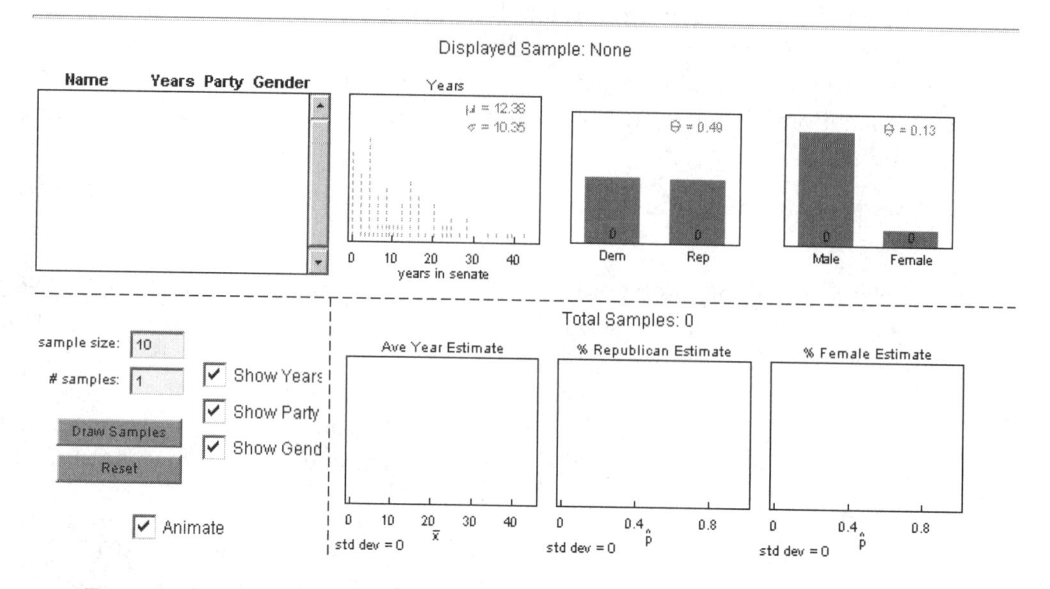

Do you think the same type of sampling bias would be present if we asked you to think of five members of the 2002 Senate that you had heard of? Explain.

c. The three graphs in the upper right part of the webpage show the distributions of how many years the senators have served, their political parties, and their genders. Are these population distributions or sample distributions? Explain.

d. The applet will allow you to take random samples from this population. Specify 10 as the sample size, and click the **DRAW SAMPLES** button. What is the average length of service in your sample? What proportion of the senators in the sample were Republican? What proportion of the senators in the sample were female?

e. Are the values reported in part d statistics or parameters?

f. Click the **DRAW SAMPLES** button again. Did you get the same values as you did in part d?

g. Change the number of samples (**# samples**) from 1 to 100, and click the **DRAW SAMPLES** button. As the computer generates samples, answer the following questions:

- Does the group of senators you obtain change from sample to sample?

- Does the average length of service you obtain change from sample to sample?

- Does the average length of service for the entire senate change from sample to sample?

- Does the proportion of Republicans change from sample to sample?

- Describe the distribution (shape, center as measured by the mean, and spread as measured by the standard deviation) of the sample averages shown in the bottom left graph (average your estimate).

- Describe the distribution (shape, center as measured by the mean, and spread as measured by the standard deviation) of the sample proportion shown in each of the other two bottom graphs (the percentage of Republicans estimate and the percentage of females estimate).

h. Now change the sample size from 10 to 5, uncheck the **ANIMATE** box, and click the **DRAW SAMPLES** button. How do the new distributions (in black) compare to the previous distributions (in green)?

Solution:

a. You may have suspected that people would have been more likely to remember the senators who have been around longer, such as Kennedy and Helms.

b. The same issue would be present in 2002, although the election of Hillary Clinton and other notable freshmen senators might make the bias less evident.

c. These three distributions are all populations, just for different variables.

d. Results will vary from student to student.

e. They are statistics since they are characteristics of samples.

f. No, the values of the statistics will vary from sample to sample.

g. Each sample will differ each time, the average length of service will change each time, and the sample proportions will change each time. However, the population parameters, $\mu = 12.38$ years and $\theta = .49$ or $\theta = .13$ do not change when we draw another sample. The distribution of the sample means should look like a mound-shaped, symmetric distribution with mean near 12.38 and standard deviation near 3.2. The distribution of the sample proportions of Republicans should look like a mound-shaped, symmetric distribution, with mean near .49 and standard deviation near .15. The distribution of the sample proportions of females will be skewed to the right, with mean near .13 and standard deviation near .1.

h. Decreasing the sample size increased the variability of the distributions. All three standard deviations are now larger than they were before. The centers of the three distributions will not change much.

Topic 13: DESIGNING STUDIES

In this topic we switched from considering how the observational units that we record information about are determined, to exactly what we do with them. Are we asking them questions? Are we forcing them into groups? Are we merely observing them? Note that it's possible to carry out these different types of studies even without a random sample. Thus, you will want to ask yourself two questions:

- Can I generalize the results to the population?

  To answer this, consider whether the observational units were *randomly selected* from the population of interest.

- Can I draw a cause-and-effect conclusion?

  To answer this, consider whether the observational units were *randomly assigned* to treatment groups.

13

Example: Suppose you want to determine whether the caffeine in a cup of regular coffee, consumed each morning before class, can improve the performance of a typical student on a statistics exam. All students in the 9 A.M. section are given the treatment (1 cup of coffee) and all students in the 8 A.M. section are not permitted to have any caffeine before class.

a. Identify the observational units, the explanatory variable, and the response variable.

b. Is this an experiment? Explain.

c. This is not a well-designed experiment. Give one reason why it will be difficult to draw conclusions about the effect of caffeine on students' exam performance from this study.

d. Suggest a design for another study of the effects of coffee on exam performance, and explain the benefits of this approach over the first study. Include a sketch outlining your study.

e. Is your design in part d double-blind? Why is this of concern?

Solution:

a. The observational units are students. The explanatory variable is whether or not the student consumes coffee (categorical), and the response variable is the student's score on the statistics exam (quantitative).

b. Because the explanatory variable (whether or not they drank coffee) was imposed on the subjects, this is considered an experiment. However, it is not a randomized experiment. Not all students had an equal chance of being assigned to drink coffee.

c. There could be other differences between the two sections besides the treatment. For example, if students in the 9 A.M. section got more sleep, that fact, rather than coffee drinking, could explain why that group did better. You have no way of separating the impact of coffee from the impact of sleep. Maybe the students who signed up for the 9 A.M. class tended to have higher motivation and better study skills and were more organized or had priority in choosing their classes than those in the 8 A.M. section. You have no way of knowing whether the improvement was due to the coffee or due to a prior difference between the two groups.

d. For a randomized comparative experiment, you would number all of the students in the two classes and use a random number table to choose half to drink coffee and half to not drink coffee as suggested in the diagram below. This random assignment will work to "equalize" the two groups. Both groups should then have about the same number of students who got enough sleep and who didn't. Both should have about the same number of highly motivated students and the same number of less motivated students. That doesn't mean that there will be an equal number of highly motivated and less motivated students in each group, but that the composition of the two groups with regard to motivation is similar.

For another approach you would number all of the 8 A.M. students and randomly assign half to drink coffee and then number all of the 9 A.M. students and randomly assign half to drink coffee. This is a type of *block* design; it allows you to maintain the distinction between the meeting time of the class in case that has an effect on test performance. This design also allows you to compare the caffeine effect among the 8 A.M. students alone, which might be useful because the caffeine effect within each class group might be more similar than between class groups.

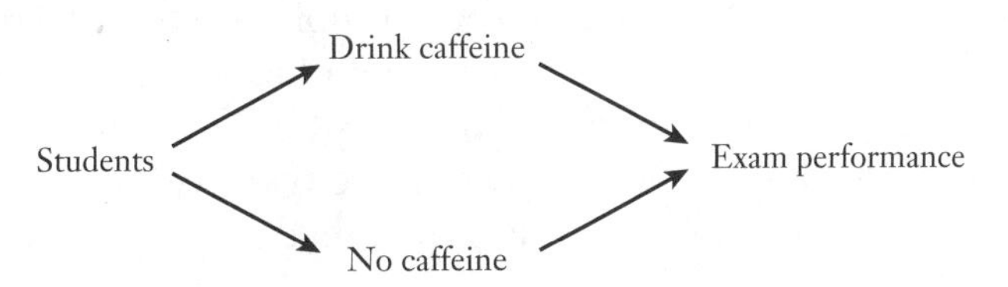

e. If some students drink coffee and some do not, then the subjects would not be blind as to whether or not they are receiving the treatment. This is especially problematic if they are told that by drinking the coffee it is expected that they will perform better on this exam. It is possible that the mere suggestion of improvement will build the students' confidence and be the cause of improved scores (the coffee drinking would be confounded with the suggestion). One way around this problem is to give caffeinated coffee to one group and decaffeinated coffee to the other group, without the students knowing which they are receiving. Similarly, the instructor reading the exam responses should not know which students had caffeine, because that could subconsciously affect how he or she grades the responses.

KEY CONCEPTS

- Randomly assigning subjects to treatment groups works to create groups that are similar except for the treatments. This method is much more reliable than using human judgment to try to equalize all possible variables between the groups.
- A random comparative experiment (with or without blocking) is the only type of study that warrants cause-and-effect conclusions (though this still does not

guarantee that the conclusion can be generalized to a larger population). By randomly assigning the subjects to groups you guard against the possibility of another systematic difference between the two groups (which would be confounded with the explanatory variable).

- An experiment is distinguished by the active imposition of the explanatory variable.
- To be a confounding variable, two things are necessary: the variable has an effect on the response variable and acts differently among the explanatory variable groups. For example, in the coffee study, "having an extra hour of sleep" could conceivably affect exam performance and could be a consistent difference between students in the 9 A.M. and the 8 A.M. sections. "Whether the students were regular coffee drinkers" could explain their reaction to the coffee and affect the response, but it does not seem as plausible that this variable would be different for students in the 8 A.M. section than for those in the 9 A.M. section. Not all lurking or extraneous variables are confounding. When describing a confounding variable, it is important to discuss how the variable differentiates the treatment groups.
- Incorporating blindness into your experiment is a good extra step for eliminating other potential confounding variables, such as the placebo effect or the inconsistent measurement of the response by the researcher.
- Incorporating a blocking variable is another extra step for further ensuring that the treatment groups are similar. If you only compare coffee effects for students in the 9 A.M. section, then you have eliminated another source of variability: class time. This makes it easier to see the effects of the coffee. The most basic type of blocking is a matched pairs design where you compare two treatments on the same individual. You can't get much more identical than that!

CALCULATION HINTS

- It is convenient, though not mandatory, to have an equal number of subjects in each experimental group.
- It is better to assign numbers to all potential participants, choose half the numbers for one group, and then assign everyone else to the other group (subjects to treatments), rather than going back and forth between groups (treatments to subjects). For example, if you flip a coin and put subjects in Group 1 if it comes up heads, once you have the right number of people in Group 1, the rest go to the other group. If, however, subjects arrived in some kind of order, there could be something distinct about "the rest," and this method could be flawed.

COMMON OVERSIGHTS

- Not explaining confounding sufficiently; for example, only talking about the effect on the response.
- Not being able to distinguish between random selection and random assignment and the purposes of each. Random selection means selecting a sample from the larger population where the goal is to be able to generalize back to that population.

Random assignment means assigning the subjects to groups where the goal is to be able to draw cause-and-effect conclusions. We don't usually refer to bias with random assignment, and we don't usually refer to confounding with random sampling. Correct usage of statistical terminology makes it much easier for the reader to be clear about your meaning.

- Not giving enough details about your experimental design. You need to specify, at least, the experimental units, the explanatory variable, the response variable, and how you plan to randomly assign the experimental units to groups. For the random assignment, you need to include enough detail that another person could carry out your plan based on your description alone (discuss the process). When stating the explanatory variable and response variable, remember to state them as variables (e.g., "whether or not drank coffee," not "whether caffeine has an effect").
- Not considering an obvious blocking factor. If one variable is likely to affect the response, block on that variable first.
- Stating a cause-and-effect conclusion when assignment to groups was not random.
- Going overboard in the other direction and thinking that cause and effect can never be established, even with randomized comparative experiments.

WHAT WENT WRONG?

The following are examples of errors in the analysis of the example problem. In each case, explain what is wrong.

1. Many early studies on the relationship between smoking and lung cancer found that smokers were about 13 times more likely to die from lung cancer than nonsmokers. Still, people argued against a cause-and-effect conclusion, citing numerous possible confounding variables. Suppose a student argues that these studies are not convincing evidence because the researchers did not record the diets of the individuals. Explain why this argument is incomplete.

2. Suppose you did a randomized comparative experiment to see whether or not caffeine helped students on a midterm by randomly assigning half the students to drink caffeine and the other half to drink a noncaffeinated coffee-flavored beverage on the morning of the exam. A student argues that you can't draw a cause-and-effect conclusion because you didn't control for what the students ate for breakfast that morning. Explain why this argument is flawed.

Solutions:

1. The student has identified a good potential confounding variable but has not yet explained why that variable is confounded with whether or not the subjects smoke. To cite this as a confounding variable, the student would need to argue that smokers have a much different diet from nonsmokers and that an unhealthy diet could be the cause of the cancer instead of smoking.

2. This student has not trusted the randomization. Because the subjects were randomly assigned to the two groups, the overall food consumption of the two groups should look similar. It may not be the same for every individual, but both groups should have a similar distribution of students who skipped breakfast, students who had a light breakfast, and students who had a hearty breakfast. Since this variable will look similar across the two groups the intake of caffeine should be the key difference between them. Therefore, if you do see a difference later between the two groups, you will have evidence that the caffeine is the cause.

FURTHER EXPLORATION

Open the "Randomization of Subjects" applet at *www.rossmanchance.com/applets/Subjects/Sub.html*, and click the **RANDOMIZE** button. You should see 12 students being shuffled and randomly assigned to two groups. The applet then graphs how many men and women are in each group.

Randomizing Subjects

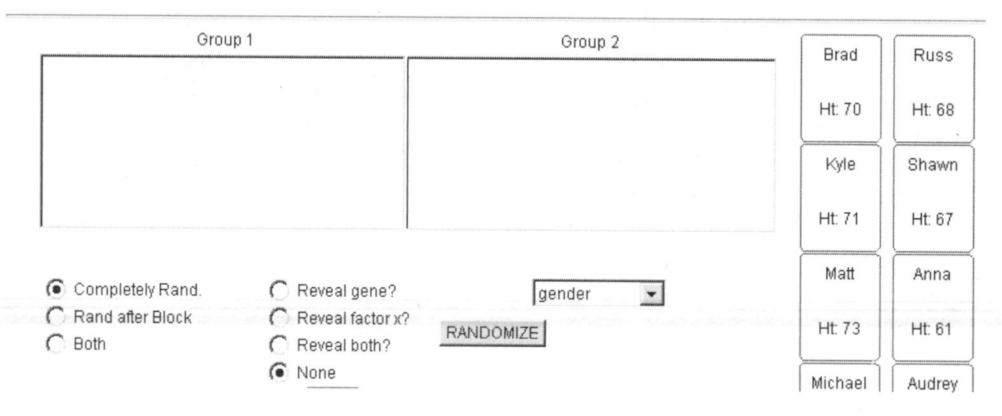

a. What proportion of Group 1 subjects were male (blue)? What proportion of Group 2 subjects were male (blue)? Calculate the difference in these two proportions.

b. Click the **RANDOMIZE** button again. Did you get the same difference in the two proportions?

c. Suppose you repeated this process many, many times. What do you conjecture will be the behavior of the distribution of these differences? For example, what do you think will be the average value of these differences?

d. Click **RESET**. Change the number of replications from 1 to 100, and uncheck the **ANIMATE** box. Click the **RANDOMIZE** button. The graph shows the difference in the two proportions for 100 repetitions of this randomization process. What is the average value of these differences? Is it close to zero? Is this what you predicted in part c?

e. Use the pull-down menu to change the variable from gender to height. The graph shows the difference in the average heights of the two groups for 100 repetitions of this randomization process. What is the average value of these differences? Is it reasonably close to zero?

f. Write a paragraph explaining how randomization evened out the gender and height variables between the two groups.

g. Would randomization also tend to create similar groups even on variables that could not be seen or measured? Explain.

Solution:

a. Answers will vary from student to student.

b. Answers will vary, but you probably got a different difference in proportions.

c. Conjectures will vary from student to student.

d. You should see that the differences are centered around zero—on average, the proportion of men in each group will be equal. Sometimes there are more men in one group than the other, but the two proportions tend to be quite similar.

e. If you examine a new variable such as height, it should also look similar in the two groups. You should see that the difference in the average height of the two groups is also around zero on average.

f. Randomization creates groups that are similar. The difference between the proportions or averages of the two groups should be around zero in the long run. You can also make predictions about how far from zero these differences can be expected to be, even though randomization was used. With these predictions you can judge whether an observed difference between two groups is large.

g. Yes, another virtue of randomization is that it works equally well on variables that you cannot easily see or measure.

Unit IV:

Randomness in Data

Topic 14: PROBABILITY

In this topic, you worked with the definition of probability. You could take an entire course that talked only about interesting applications of probability, but we focused on the basic principles only and started to introduce how these ideas are used in statistics. The question "How often does this happen in the long run?" will be of key interest throughout the rest of the course. One way to learn about probabilities is to simulate the random process many times to help give you intuition as to whether certain outcomes are likely or not.

Example: Many fast food restaurants run games where you need to collect cards until you have a complete set. Suppose one chain wants you to collect railroads until you have all four (you have to keep purchasing cards until you have collected at least one of each card). Suppose further that 40% of the cards are Railroad A, 25% are Railroad B, 25% are Railroad C, and 10% are Railroad D.

a. If you select a card at random, what is the probability that you will select a Railroad A card?

b. Provide an interpretation of the probability calculated in part a.

c. Repeat part a for the Railroad B, C, and D cards. What should these probabilities sum to?

14

d. Explain how you could use the random number table to simulate playing this game until you have a complete set of the four railroad cards, determining how many cards you need to buy.

e. Conduct the simulation in part d ten times. Use your simulation results to provide an estimate of the expected number of cards you need to buy in order to obtain a complete set.

Solution:

a. If 40% of the cards produced are for Railroad A, and the cards are truly distributed at random, then the probability is .40 that you will randomly select a Railroad A card.

b. This probability means that if you were to repeatedly draw cards forever and examine the proportion of times you obtain a Railroad A card, then this proportion will approach .40. In other words, if you were to buy a card many, many times, then in the long run 40% of the cards you bought would be Railroad A cards.

c. Railroad B: .25, Railroad C: .25, Railroad D: .10

Since these are the only four possible cards, these probabilities must sum to 1.

d. To match the stated proportions, use two-digit numbers. One approach is to let the numbers 00–39 represent a Railroad A card, 40–64 represent a Railroad B card, 65–89 represent a Railroad C card, and 90–99 represent a Railroad D card. Then select a pair of digits and keep track of which railroad card it represents. Continue until you have at least one tally mark by each card. Then report how many cards you had to "buy." For example, using line 48 of Table I:

78892 99777 18049 66117 78028 70955 75476 98203 01512 23591

you would get the following ID numbers:

78, 89, 29, 97, 77, 18, 04, 96, 61

corresponding to the following cards:

C C A D C A A D B.

You needed to buy nine cards in order to obtain each railroad at least once.

e. If you take the average of the number of cards needed to complete a set, this gives an estimate of the expected number of cards. The more you repeat this process, the closer you get to the true long-run average: the expected value. However, with only ten repetitions, your estimate may not be all that close.

KEY CONCEPTS

- Probability is interpreted as the long-run relative frequency of an outcome. You may need to observe a random event for a rather long time before obtaining a precise estimate of the limiting value of the relative frequency.
- You should not be too surprised by short-run deviations from the theoretical probability. For example, if you flip a coin four times, you are actually more likely *not* to get two heads and two tails than you are to get the even split.
- In this course, all probabilities will be expressed as decimals or fractions and will always be between 0 and 1, inclusive. If you sum the probabilities of all possible outcomes, they will sum to 1.
- You can estimate probabilities through simulation or calculate probabilities theoretically. For example, if the outcomes are equally likely, you can determine the theoretical probabilities by listing all possible outcomes and counting how many have a certain property.
- In many situations, not all outcomes are equally likely.
- Unfortunately, common sense can be pretty faulty when it comes to determining probabilities. In other words, some concepts related to probability are counter-intuitive. Don't rely on your intuition. Carefully follow the probabilities rules that are presented.
- Sample size plays an important role in probability calculations. The larger the sample size, the more likely is an expected event and the less likely is an unusual event. For example, when a fair coin is flipped 100 times, the proportion of heads is more

likely to be close to .5 (say, between .45 and .55) than when it is flipped 10 times. On the other hand, obtaining a sample proportion of heads greater than .75 is more likely with 10 flips than with 100 flips.

CALCULATION HINTS

- You learned that one way to calculate probabilities is to count the number of times an outcome occurs and divide by the total number of possible outcomes. Keep in mind that this technique only applies when the outcomes are equally likely.

COMMON OVERSIGHTS

- Assuming that probabilities hold in the short run. For example, if a basketball player misses three free throws in a row, people often conclude there is a problem. However, even an 80% free throw shooter will occasionally have three misses in a row, just by chance. Saying a player's probability of making a free throw is .8 means that if you were to observe the player's performance thousands and thousands of times, in the long run, roughly 80% of the shots would be successful.

WHAT WENT WRONG?

The following are examples of errors in the analysis of the example problem. In each case, explain what is wrong.

1. The following is a transcript from a *Nightline* interview conducted by Ted Koppel on October 13, 1997 (transcript found at *www.dartmouth.edu/~chance/chance_news/recent_news/chance_news_6.12.html#Everyprobability*). Critique the earthquake expert's probability statements.

   TED KOPPEL: *Dr. Andrews, I'm sure you have heard such cautionary advice before, so on what basis is the assumption being made that this is the one that's going to have the kind of impact on southern California in particular that's being predicted?*

   RICHARD ANDREWS: Well, in the business that I'm in and that local government and state government is in, which is to protect lives and property, we have to take these forecasts very seriously. We have a lot of forecasts about natural hazards in California, and we have a lot of natural events here that remind us that we need to take these forecasts seriously. I listen to earth scientists talk about earthquake probabilities a lot and in my mind every probability is 50-50, either it will happen or it won't happen. And so we're trying to take the past historical record, our own recent experience of the last two of the last three years and make the necessary preparedness measures that can help protect us as much as we can from these events.

2. Identify what's wrong with all of the following arguments:

   a. Suppose that Andre plays a video game for which his probability of winning is .4, and he plays for a total of 80 turns. If he wins on at least half of his turns, he comes out ahead in the game. Andre argues that he has a .4 probability of coming out ahead.

   b. Suppose that Emily plays the same video game and has the same .4 probability of winning on every turn, but she only plays for a total of 20 turns. Andre contends that Emily has the same probability of coming out ahead as he does.

   c. Suppose that Larry makes 80% of his free throws and Tim makes 60% of his free throws. Larry is more likely than Tim to make all of his free throws.

   d. In a family of two children, the probability is 1/3 that there will be two boys, 1/3 that there will be two girls, and 1/3 that there will be one of each gender.

   e. Suppose that two candidates are to be hired from a pool of 40 applicants. Suppose that three of the applicants are related to the supervisor, and it turns out that both people hired are his relatives. The supervisor claims that this is not surprising because there were only three possible outcomes: that two of his relatives would be hired, that one would be, or that none would be. Thus, there was a 1/3 probability that both of his relatives would be hired, and this does not constitute a surprising outcome.

Solutions:

1. The interviewee is considering a variable—whether or not there is an earthquake—and has assigned probabilities to the two possible outcomes (yes, no) that sum to 1 (.5, .5). It seems unlikely that he really intended Pr(*earthquake*) = .5. This would mean that if he took part in a random process where he recorded whether or not there was an earthquake many, many times, then in the long run, 50% of those observations would be an earthquake. It is not clear whether this expert is referring to a specific location or specific time period (year? day?), but he seems to be assuming that simply because there are only two possibilities, they must be *equally likely*. It is important to not assume equal likeliness. Even more egregious examples are to assume that because there is either intelligent life on Mars or not, the probability of intelligent life is .5, or that because a nuclear holocaust will either occur within 5 seconds of your reading this sentence or not, the probability of a nuclear holocaust is .5.

2. **a.** The .4 probability means that, in the long run, Andre will win about 40% of his turns at the game. This probability applies to individual games, not to a large number of games. If he plays 80 turns, the proportion that he wins is more likely to be close to this .4 probability than far from it, so the probability that he wins more than 50% of his turns (and comes out ahead) is fairly small, much less than .4. It can be shown that this probability is about .044.

   b. Since she plays fewer turns than Andre, Emily is less likely to see her proportion of winning turns fall close to the long-run probability of .4. So, Emily is more likely to win on at least half of her turns than Andre is. You could convince yourself of this through simulation or by making the argument more extreme by considering a sample of two turns only. The probability that Emily wins on at least one of her two turns can be shown to be about .245, which is much higher than Andre's.

   c. This is correct if Larry and Tim attempt the same number of free throws per game. But if, for example, Larry attempts 10 free throws per game and Tim attempts 2, then Tim would be more likely to make them all despite his lower probability of success.

   d. This is a misapplication of the equal likeliness assumption. There are indeed only three outcomes—two boys, two girls, and one of each—but the "one of each" outcome is twice as likely as the others because it can happen in two ways: first child boy and second girl, or vice versa. If we write the sample space as B_1B_2, B_1G_2, G_1B_2, and G_1G_2, then these four outcomes are equally likely, with probability .25. Thus, the probability of getting one child of each gender is .5.

   e. The supervisor is misapplying the concept of equal likeliness in order to deflect attention from how surprising this result would be if the hirings were made at random. The three outcomes that he lists—that two of his relatives would be hired, that one would be, and that none would be—are not equally likely because there are so many more nonrelatives who applied. The actual probability that both hirees would have been among the supervisor's three relatives who applied can be shown to be around .004, which of course is much less than 1/3. So, this outcome would be very surprising if the relatives did not have an advantage in the hiring process.

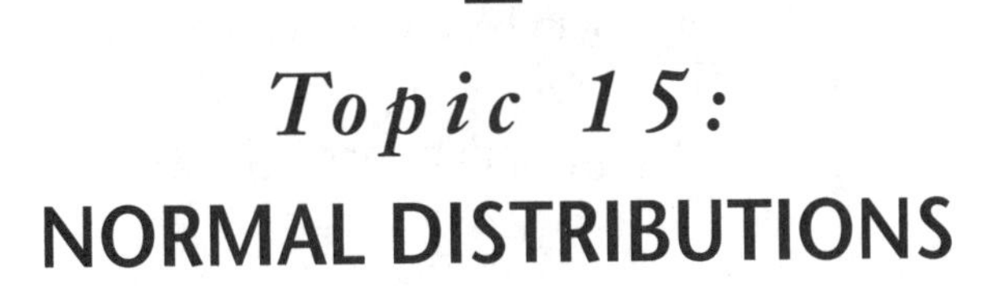

Topic 15:
NORMAL DISTRIBUTIONS

In this topic you worked with a very specific, very common probability model—the normal distribution. You should have been convinced that the normal distribution comes up in many different types of situations. Calculating probabilities from a normal distribution is straightforward as long as you start with a sketch and follow a series of steps to determine the probability of interest.

Example: The following histogram shows the results of the measurement of the index finger length (in centimeters) for 3000 criminals.

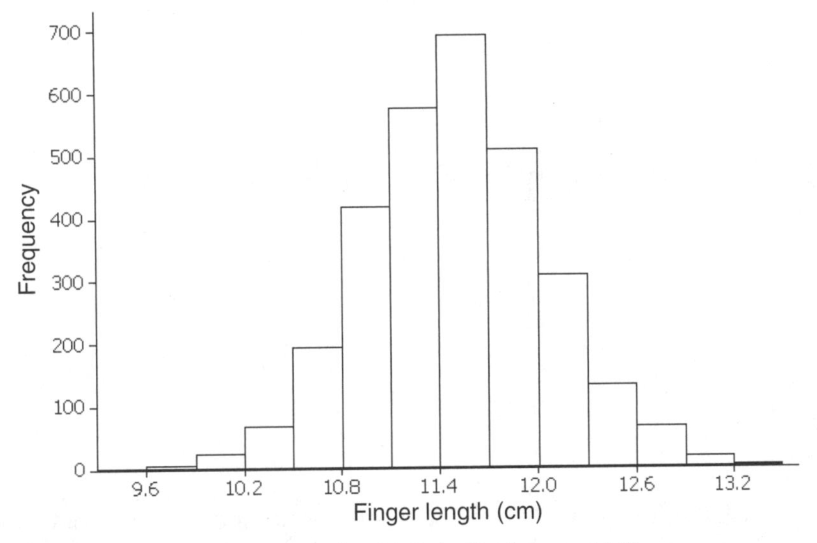

Source: Handbook of Small Data Sets by Daly, Hand, Lunn, McConway, Ostrowski (Editor) (CRC Press, Boca Raton: 1993)

a. Do these data appear to follow a normal distribution? Explain.

b. From this histogram, provide a rough estimate of the mean and standard deviation of this distribution.

15

Suppose you believe that the index finger lengths of criminals follow a normal distribution with mean $\mu = 11.5$ cm and standard deviation $\sigma = 0.55$ cm (these estimates are based on the preceding distribution).

c. Where do you expect to find the middle 95% of these lengths?

d. What percentage of criminals do you expect to have a finger length less than 12 cm?

e. How long are the longest 5% of criminal index finger lengths?

Solution:

a. This distribution is symmetric, with one peak, and does not have many very extreme values, so, yes, the data do appear to follow a normal distribution.

b. The center of the distribution appears to be around 11.5 cm, so that is a reasonable estimate of the mean. The edges of the middle 2/3 of the observations appear to be roughly 11 and 12 cm, giving a standard deviation of 0.5 cm.

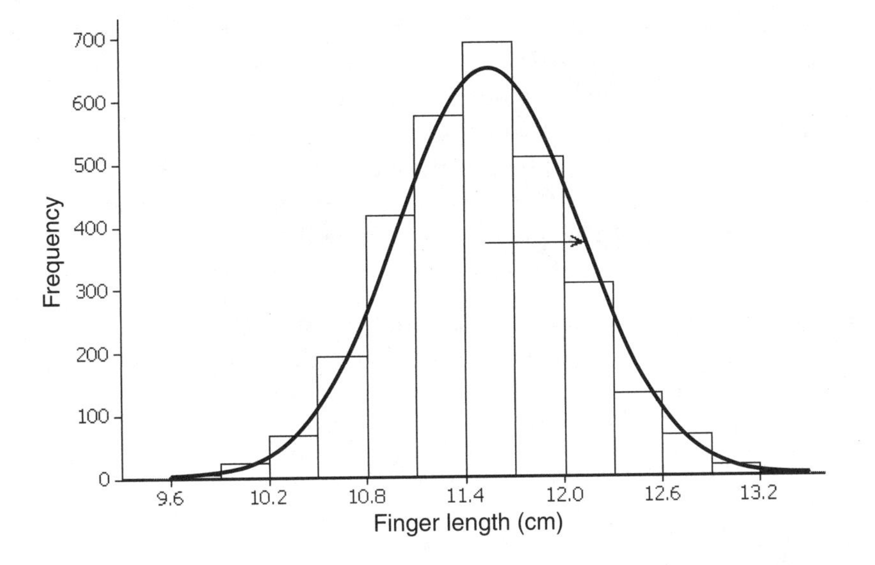

c. With a mean of 11.5 and a standard deviation of 0.55, the middle 95% of these observations should fall within two standard deviations of the mean: $2(0.55) = 1.1$.

Two standard deviations below the mean $= 11.5 - 1.1 = 10.4$

Two standard deviations above the mean $= 11.5 + 1.1 = 12.6$

So you would expect the middle 95% of the finger lengths to be between 10.4 cm and 12.6 cm.

d. Working with the normal (11.5, 0.55) distribution, start with a sketch of the distribution and shade the region of interest.

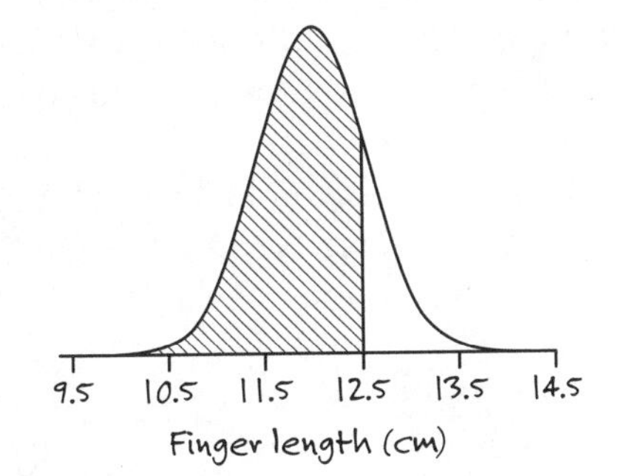

The question asked about lengths *less than* 12 cm. Since 12 cm is larger than the mean, this probability must be larger than .5. To determine this probability exactly, *standardize* the observation:

$$z = \frac{\text{observation} - \text{mean}}{\text{standard deviation}} = \frac{12 - 11.5}{0.55} = 0.91$$

This z-score tells us that a finger length of 12 cm is almost one standard deviation above the mean of the finger lengths.

15

If you look up the value 0.91 in *WS* Table II, you will see that the probability below $z = 0.91$ (i.e. $\Pr(z \leq 0.91)$ is equal to .8186. Therefore, the probability of selecting a criminal index finger length less than 12 cm is .8186. Alternatively, you could state that approximately 82% of criminal index finger lengths are less than 12 cm. This approximation is based on the normal model. For the original histogram, you can count 2481 lengths that are at most 12 cm. This corresponds to 82%, giving evidence that the normal distribution describes these finger lengths reasonably well.

e. Again, using the normal distribution, for an index finger length to be in the top 5% you need to find a length such that 95% of the lengths lie below that value. Start with a sketch:

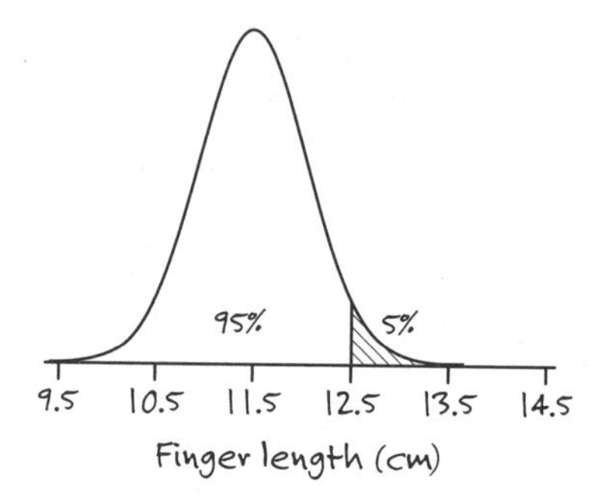

Since the probability .95 is given, you next convert this probability to a z-score. Using the probabilities in Table II, look for the four-digit number as close to .9500 as you can find. You will find .9452 and .9554. These correspond to z-values of 1.64 and 1.65, respectively. So the z-value is somewhere between these two numbers. If you take $z = 1.64$ (it's slightly closer to that value since .9452 is a little closer than .9554 to .95), this means that to be in the top 5% of criminal index finger lengths, the length needs to be at least 1.64 standard deviations above the mean. So now you can convert this z-value into an actual finger length. To be 1.64 standard deviations above the mean, the value needs to be $1.64(0.55) = 0.902$ above 11.5 or $11.5 + 0.902 \approx 12.4$. Thus, finger lengths of 12.4 cm and greater are in the top 5% of this distribution. Another way to see this last step is to look at the formula for the z-score:

$$z = \frac{\text{observation} - \text{mean}}{\text{standard deviation}},$$

where you know $z = 1.64$, the mean, and the standard deviation. You can solve algebraically for the observation:

$$1.64 = \frac{\text{observation} - 11.5}{0.55}$$

$$0.55(1.64) = \text{observation} - 11.5$$

$$0.55(1.64) + 11.5 = \text{observation}.$$

As a reality check, make sure that the observation value you determined is greater than the mean, because you want the "top" 5%.

KEY CONCEPTS

- The normal distribution is a very powerful model for describing the behavior of many quantitative variables.
- The empirical rule applies very well for normal distributions. From the empirical rule, you know that 95% of observations should be within two standard deviations of the mean.
- The z-score, or standard score, is the "ruler" used for measuring distance. The z-score indicates how many standard deviations fall between the observation and the mean of the distribution. The larger the standard deviation, the smaller the z-score, which also affects the probabilities you calculate.
- Using the z-score and Table II, you can determine the probability of the occurrence of certain outcomes in normal distributions. These probabilities are represented by the area under the normal curve in the region of interest and also indicate the probability that a randomly selected observation will have the characteristic in question (e.g., low birth weight).
- Using the z-score and Table II, you can determine which values or regions correspond to stated probabilities or percentiles.
- The virtue of the z-score is that it standardizes observations by taking into account the mean and standard deviation. This enables you to compare values that may come from different distributions. For example, suppose you had measured finger length in inches instead of centimeters and wanted to use the normal distribution with mean 4.5 inches and standard deviation 0.22 inches to model these finger lengths:

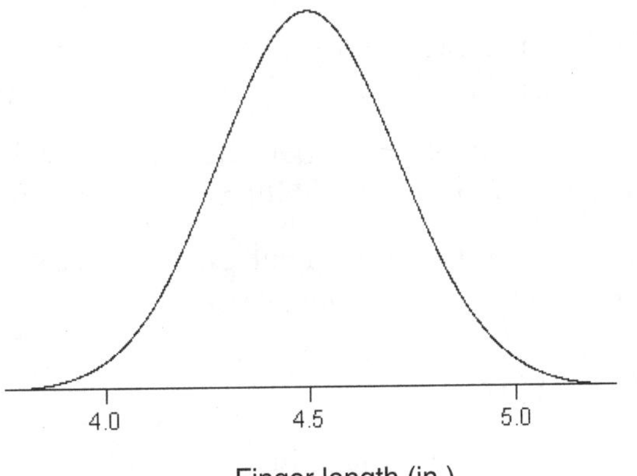

In the example you looked at a finger length of 12 cm. Since there are 2.54 cm in 1 inch, this corresponds to a finger of length 4.7 inches. This observation, 4.7 in., should have the same *relative position* in the distribution as 12 cm did in the original model. In fact, if you standardize this value: $z = (4.7 - 4.5)/.22 = .91$, you find the exact same z-value. Even though the graph is scaled differently, the z-scores enable you to put observations from different distributions onto the same "standard" scale.

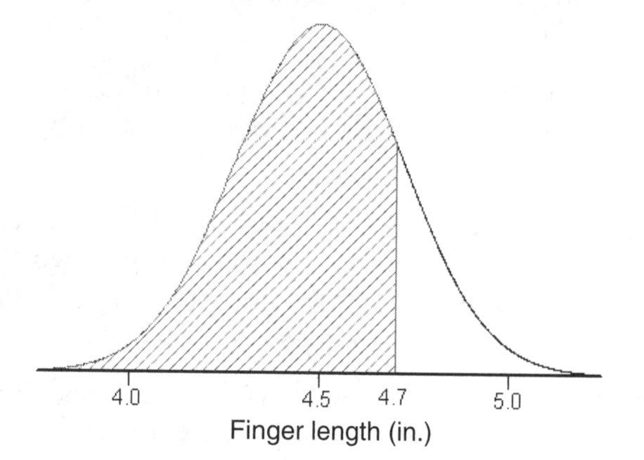

CALCULATION HINTS

- Start with a sketch of a normal distribution! Label and scale the horizontal axis, indicating the mean and at least one standard deviation to either side, and shade the probability of interest.
- First ask yourself if you were given an observation and asked to find the probability, or given the probability and asked to find the observation.
- Avoid sloppy notation, such as $z = .91 = .8186$. Two different statements have been collapsed into one here. The z-score is $z = 0.91$, and the probability below this z-score is .8186.
- If you are looking for the probability that the outcome is greater than a certain value, you must subtract the probability in *WS* Table II from 1.
- If you are looking for the probability that the outcome is between two values, you must subtract the corresponding probabilities found in Table II.
- Do a reality check at the end of your calculation to make sure the answer you found makes sense and is consistent with your sketch.

COMMON OVERSIGHTS

- Confusing the z-value with the probability.
- Not subtracting from 1 when asked to find the probability that the outcome is greater than a specific value.
- Subtracting the z-values when asked to calculate the probability that the outcome is between two specific values. You must first convert the z-values to probabilities, which can then be subtracted.

WHAT WENT WRONG?

The following are examples of errors in the analysis of the example problem. In each case, explain what is wrong.

Assuming criminal finger lengths follow a normal distribution with mean 11.5 cm and standard deviation 0.55 cm, a student was asked to find the probability that a criminal finger length is:

1. Shorter than 11.6 cm
2. Longer than 11.6 cm
3. Between 12 and 12.5 cm
4. Longer than 12 cm

Her work is shown next. What mistakes has she made?

1. $\frac{11.6 - 11.5}{0.55} = 0.18$

   There are 18% less than 11.6.

2. $\frac{11.6 - 11.5}{0.55} = 0.18$

   From Table II, the probability is .5714.

3. $\frac{12 - 11.5}{0.55} = 0.91$ $\frac{12.5 - 11.5}{0.55} = 1.82$ $1.82 - 0.91 = 0.91$

   The probability is .91.

4. $z = \dfrac{12 - 11.5}{0.55} = 0.91$

$1 - .91 = .09$

From Table II, the probability is .5359.

Solutions:

1. This student should start with a sketch of the normal curve, shading in the probability of interest.

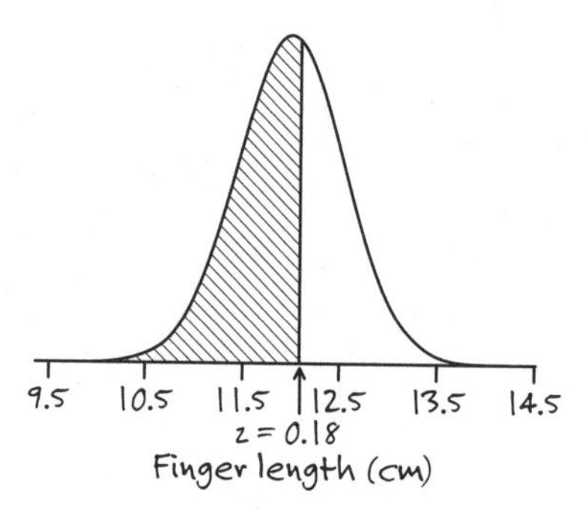

Then the student should label the calculation as being that of the z-score. Both of these steps should have reminded the student that the z-score has been calculated, but the next step is to use Table II to convert this z-score into a probability. This probability turns out to be .57, which is much more consistent with the sketch. In particular, you know the probability has to be larger than .5 because the observation of interest was larger than the mean. This student also failed in all four answers to state the solution in context—18% *of what*? Make sure it is clear from your final statement what the observational units and variable are, for example, "57% of criminal finger lengths are less than 11.6 cm."

2. This time the student did look up the z-score in Table II to find a probability. But the probability reported is the probability that a finger length is *less than* 11.6 cm. She needs to subtract from 1 in order to find the probability that a finger length is

longer than 11.6 cm: 1 − .5714 = .4286, so about 43% of finger lengths are longer than 11.6 cm.

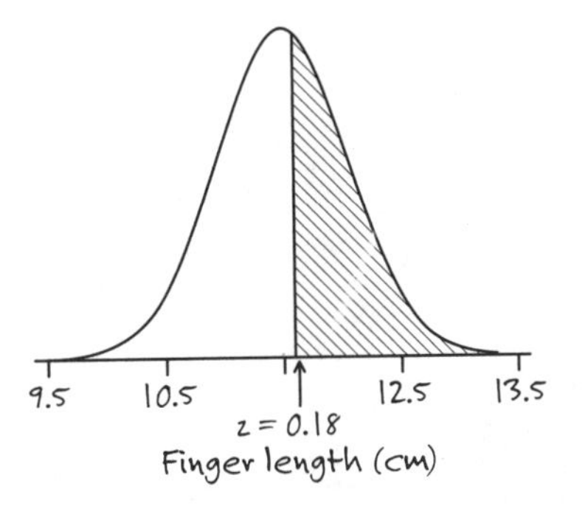

3. The z-scores are calculated correctly, but it's their *probabilities* that must be subtracted, not the z-scores themselves. Table II reveals these probabilities to be .9656 and .8186; .9656 − .8186 = .1470, so about 15% of the finger lengths are between 12 and 12.5 cm.

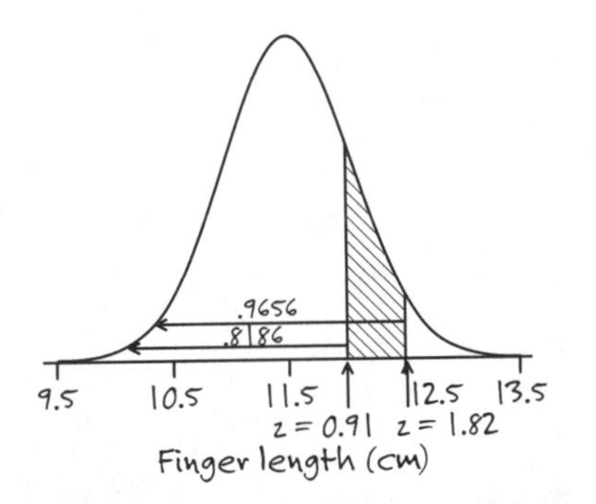

4. The student has remembered that when calculating the probability that the outcome is greater than a certain value, something is subtracted from 1. However, she must convert to a probability in Table II *first* (.8186) and then subtract the probability from 1 (1 − .8186 = .1814), not the z-score.

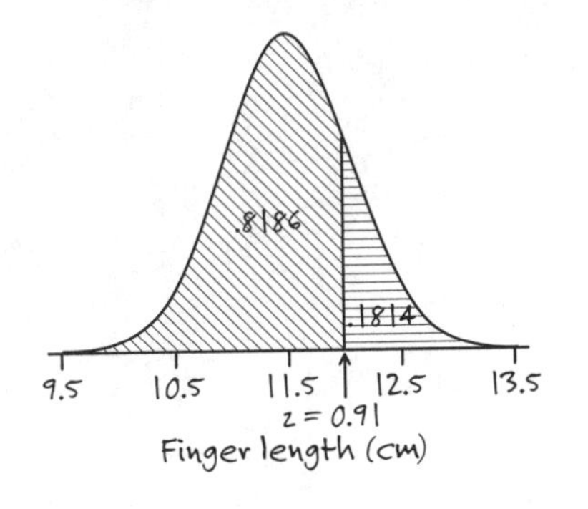

15

Topic 16:

SAMPLING DISTRIBUTIONS I: PROPORTIONS

In this topic you studied how a sample proportion varies from sample to sample and you saw that this variation is not haphazard but rather follows a predictable pattern in the long run. In fact, when the sample size is large, that pattern is a normal curve, as you studied in the previous topic. This pattern is called a *sampling distribution*, a concept that underlies all of statistical inference and the remainder of *Workshop Statistics.* Understanding sampling distributions is the key to understanding the two fundamental concepts of statistical inference: confidence and significance, which you also began to examine in this topic.

This topic involved an important transition: Instead of focusing on the distribution of individual measurements (for example, colors of candies), the focus is now on the distribution of sample proportions and how these proportions vary from sample to sample. In these distributions you can view the sample proportion as the variable and the samples as the observational units. In this topic you found that the normal distribution provides a good model of the pattern taken on by sample proportions in many random samples from the same population. Once you know the mean and standard deviation of this sampling distribution, you can apply the empirical rule and other properties of normal distributions to make statements about the expected behavior of sample proportions. Thus, in the future, if you can only observe one sample proportion, you can still make probability statements about the value of that proportion.

Example: Suppose that 30% of American adults report sleeping less than seven hours each night.

a. If American adults are the observational units, what is the underlying variable of interest here? Is it a quantitative or a categorical variable?

16

b. Is 30% a parameter or a statistic? Explain.

c. If you took a random sample of 100 American adults, would you expect to find exactly 30 of them to report sleeping less than seven hours each night? Explain.

Suppose you took 1000 different samples, each with 100 American adults, from this population. The following histogram displays the simulated sample proportions for these samples.

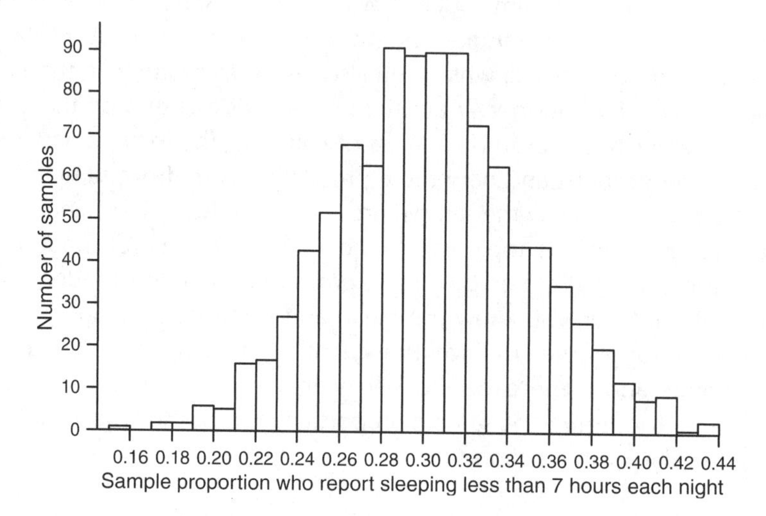

d. What are the observational units in this histogram? What is the variable? Is it quantitative or categorical?

e. About what percentage of these 1000 samples produced a sample proportion within ±.09 of the population proportion of .3?

f. Are the conditions necessary for the Central Limit Theorem for a sample proportion satisfied here? Explain.

g. Applying the Central Limit Theorem, draw a sketch of the sampling distribution of the sample proportions $\hat{p}$ in many random samples of size 100 from the population when $\theta = .30$. Make sure you label the horizontal axis and indicate the mean and standard deviation of the distribution.

h. Based on the empirical rule, does your answer to part e match what you would have predicted? Explain.

i. Describe how the preceding histogram and your answer to part e would change if the sample size had been 50 instead of 100.

j. Describe how the preceding histogram would change if the population proportion who claim to sleep less than seven hours each night had been .40 instead of .30.

Solution:

a. The observational units are adult Americans, and the variable of interest is their response to the question "Do you get less than seven hours of sleep each night?," which is a categorical variable.

b. The 30% is the percentage of the population, so it is a parameter. We use the Greek letter θ to refer to this population proportion.

c. Not necessarily; you would expect sampling variability, so the sample proportion of American adults who say they sleep less than seven hours a night might not exactly equal the population proportion .30.

d. In this histogram, the observational units are the *samples* (not the individual people anymore), and the variable that you have measured on each sample is the proportion who report sleeping less than seven hours each night. We use the symbol $\hat{p}$ to refer to this sample proportion. The sample proportion is considered a quantitative variable.

e. It looks as though about 95% of the sample proportions are between .21 and .39.

f. Yes, because $n\theta = 100(.3) = 30 \geq 10$ and $n(1 - \theta) = 100(.7) = 70 \geq 10$. Since both of these values are greater than 10, the Central Limit Theorem is considered valid here.

g. The Central Limit Theorem indicates that the sampling distribution of the sample proportion $\hat{p}$ is approximately normal with mean θ, which equals .30 here, and standard deviation $\sqrt{\theta(1 - \theta)/n} = \sqrt{(.3)(.7)/100} = .0458$.

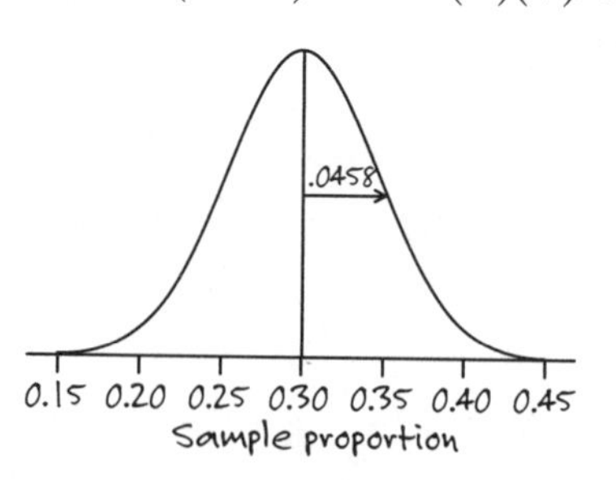

h. From part g, you can see that .09 is approximately equal to two standard deviations ($.09 \approx .0458 \times 2$). Therefore, the empirical rule states that 95% of the observations should fall within this distance from the mean of the distribution. That is, 95% of sample proportions should fall within ±.09 of .30. This agrees nicely with what is shown in the histogram.

i. The smaller sample size would produce sample proportions that are more spread out from the population proportion. The distribution would still be roughly normal and centered around .30. Fewer of the sample proportions would fall within ±.09 of the .30 since the standard deviation would now be larger.

j. The distribution of sample proportions would shift to the right, centered around .40. The variability would also increase slightly, because the new standard deviation will be $\sqrt{.4(.6)/100} = .0490$.

KEY CONCEPTS

- A parameter refers to the population, a statistic refers to a sample. A parameter is a number, but you typically do not know its actual value because it is usually not possible to gather data about the entire population.
- The value of a sample proportion varies from sample to sample (sampling variability). You can consider it a variable with the individual samples as the observational units.
- As long as you take random samples, the distribution of the sample proportions (also referred to as the sampling distribution) will follow a predictable pattern: If the sample size is large enough, the shape will be approximately normal, the mean will be equal to the value of the population proportion θ, and the standard deviation will be equal to $\sqrt{\theta(1 - \theta)/n}$.
- This pattern enables you to apply the properties of the normal distribution from Topic 15. Since you expect the sampling distribution to be mound-shaped and symmetric, you can apply the empirical rule to predict how far the sample proportions are likely to fall from the population proportion. In fact, you expect about 95% of the sample proportions to fall within two standard deviations of the population proportion θ.
- The sampling distribution becomes less variable as you increase the sample size. (Note that the sample size, n, is in the denominator of the standard deviation formula, which means that as n increases the standard deviation decreases).

16

COMMON OVERSIGHTS

- Not clearly distinguishing between a parameter and a statistic. A parameter is a specific, but typically unknown, value describing the population. A statistic is calculated from a sample and therefore varies from sample to sample.
- Trying to use the formulas given here inappropriately; for example, not checking the sample size or random sampling conditions or trying to use the formulas with a quantitative variable. These formulas should only be used for a binary categorical variable (e.g., whether or not the subject got enough sleep, whether or not the subject received an orange candy) with a large sample size.

- Confusing the different "levels." For example, the original variable may be categorical (whether or not the subject received an orange candy), but the variable of the sampling distribution (proportion of orange candies in the sample) is quantitative. The observational units for the first variable are the candies, but for the second variable they are the random samples.
- Thinking that the population size is relevant in determining the distribution of the sample proportions. As long as the population is much larger than the sample (e.g., more than 20 times as big, which will be true of the populations of all the problems in this book), then the population size is not a factor. All that matters is the sample size; we don't even care what percentage of the population we have sampled. For example, suppose your friends Julia and Emeril each make the same soup recipe. Julia makes three cups of soup and Emeril makes a restaurant-sized vat. As long as the soups are well stirred, you should be able to judge the taste equally well from one spoonful of either person's soup.
- Confusing the sample size and the number of samples. It is very easy to mix up these values and to not use language precise enough for your meaning to be clear. The sample size is how many observations are in each sample (e.g., 25 candies) and the number of samples is how many times you took a sample from the population and calculated the sample proportion. The sample size is the important value; the number of samples is just some large number, hopefully large enough for you to develop an idea of how the sampling distribution behaves. In fact, now that you have the Central Limit Theorem to predict how the distribution of the sample proportions will behave, you no longer need to simulate taking many samples from the population. So when you observe the actual sample proportion in a study, you will use the Central Limit Theorem to help you make judgments about that sample proportion and how it compares to the population proportion.
- Not recognizing that the sampling distribution of a sample proportion refers to an entire distribution, meaning that it encompasses features of shape, center, and spread.

CALCULATION HINTS

- In describing the mean discussed in the Central Limit Theorem for a Sample Proportion, it can be confusing to use the phrase "the mean is the proportion." The mean of the sampling distribution will be equal to a number. That number is equal to the value of the population proportion. It makes sense that the values of the sample proportions will cluster around the value of the population proportion. If you were to average all of the individual sample proportion values, they would average out to the value of the population proportion. A better way to phrase this relation, then, is the mean of the sample proportions turns out to be the population proportion.
- In calculating the standard deviation described by the Central Limit Theorem, remember to express the population proportion as a decimal between 0 and 1. It can be tempting to try to plug in the percentage (e.g., 30%), but that form becomes problematic when you try to subtract from 1 (e.g., 1 − 30%). This expression results in an incorrect negative number, which is especially problematic when you try to take the square root. Note that the standard deviation of sample proportions will also always be between 0 and 1.

- In checking the conditions of the Central Limit Theorem, make sure you check both of the sample size conditions, $n\theta \geq 10$ and $n(1 - \theta) \geq 10$. Always show the numbers plugged in to the formulas, the results of the multiplication, and your evaluation of whether the result is larger than 10.

FURTHER EXPLORATION

Open the "Reese's Pieces" Java applet at *www.rossmanchance.com/applets/Reeses/ReesesPieces.html.* You will see a representation of the scenario from *WS* Topic 16 in which you take random samples from a population of Reese's Pieces candies and keep track of the proportion of orange candies in each sample. One key difference here is that you will need to tell the computer what to use for the value of the population proportion θ. The following questions are intended to reinforce the observations you have made in *WS* Topic 16.

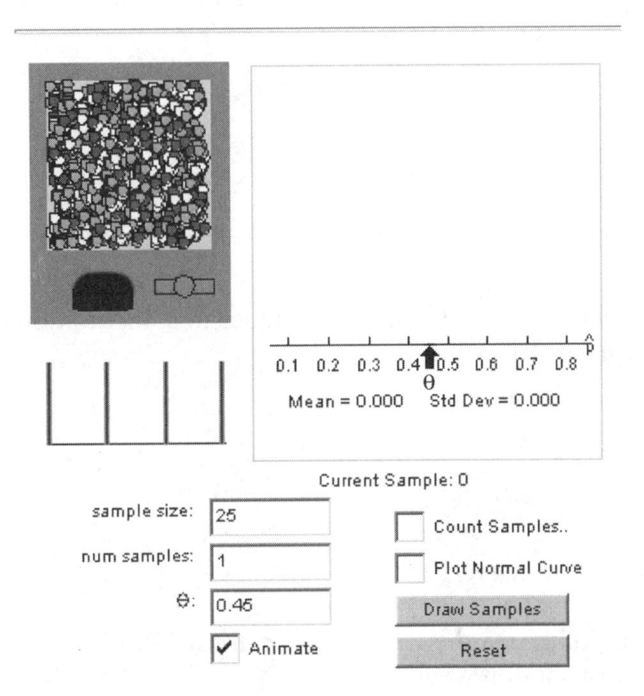

a. The value of θ has been initially specified to be .45. Click the **DRAW SAMPLES** button. What did you obtain for the sample proportion? What would be an appropriate graphical display of these sample results?

b. Click the **DRAW SAMPLES** button again. Did you obtain the same sample proportion both times? Did you expect to?

The two sample proportions have been plotted on the dotplot in the upper right. Below the dotplot, you should see the average of the two sample proportions you obtained and the standard deviation of the two values. The red arrow indicates the value of θ.

c. Click the **RESET** button, then uncheck the **ANIMATE** box. Change the number of samples from 1 to 500. Click the **DRAW SAMPLES** button. Describe the shape, center, and spread of the resulting distribution. How many sample proportions are represented in this distribution?

d. Check the **PLOT NORMAL CURVE** box. Does this distribution appear to behave like a normal distribution?

e. Suppose you changed the sample size, n, from 25 to 50. How do you think the sampling distribution would change?

f. Make the change discussed in part e, and click the **DRAW SAMPLES** button. The new distribution is represented by black dots (the previous distribution with $n = 25$ is now gray). Was your prediction correct? If not, explain why.

g. Suppose you were to change the value of θ from .45 to .70. How do you think the sampling distribution would change?

h. Make the change discussed in part g, and click the **DRAW SAMPLES** button. Was your prediction correct? If not, explain why.

i. Suppose you changed the value of n from 50 to 10. How do you think the sampling distribution would change?

j. Make the change discussed in part i, and click the **DRAW SAMPLES** button. Was your prediction correct? If not, explain why.

Solution:

a. If you look at a single sample, the variable is "whether the candy is orange," and the observational units are the candies. This variable is categorical. You could make a bar graph that displays one bar whose height equals the number of orange candies and then a second bar whose height equals the number of non-orange candies. The sample proportion obtained is the *statistic* describing this sample.

b. You probably obtained two different sample proportions, reminding you of sampling variability. It's possible to get the same value twice, but, in general, the value of the sample proportion varies from sample to sample.

c. There should be 500 values represented in this distribution, corresponding to the 500 different sample proportions. The shape of these sample proportions should be symmetric with center around .45 and standard deviation around $\sqrt{.45(.55)/25} = .099$. Make sure you used the values of the mean and standard deviation reported by the applet to support your discussion. Remember, this is an *empirical sampling distribution* based on only 500 samples, so your results may not match these values exactly (they would if you took all possible samples of size 25 from this population), but they should be within about .001 or .002.

d. The normal curve should do a pretty good job of summarizing the behavior of these sample proportions. You verified in class that the sample proportions follow the empirical rule, as you would expect for a variable that is following the normal distribution model.

e. If you increase the sample size, you expect the variability to decrease and the sample proportions to cluster more closely around .45. The center will still be .45, and the shape will still be approximately normal (if not more so).

f. You should have found the mean reported by the applet to be about .45 and the standard deviation to be about .07. [*Note*: $\sqrt{.45(.55)/50} = .0703$.]

g. Changing the value of θ will change both the center and the spread. If θ is now .70, then the distribution of sample proportions will center around .70. The standard deviation will change to $\sqrt{.70(.30)/50} = .065$, if you kept n equal to 50, or $\sqrt{.70(.30)/25} = .092$, if you changed n back to 25.

h. You should see the distribution shift to the right and become slightly less variable, but retain its normal shape.

i. With such a small sample size and $\theta = .70$, you should see that the normal shape does not describe the distribution nearly as well. The distribution is much more granular and will have more of a skewed shape.

j. The sampling distribution skews slightly to the left and is rather granular. For these reasons the distribution is not well described by the normal curve. If you aren't convinced yet, try a sample size of $n = 5$ or make the value of θ even closer to 1, perhaps with a smaller number of samples.

WHAT WENT WRONG?

The following are examples of errors in the analysis of the example problem. In each case, explain what is wrong.

1. A student was given this question: Researchers indicate that about 20% of college students abstain from drinking alcohol each year. Suppose you took a random sample of 85 college students from across the nation and calculated $\hat{p}$ (the proportion who say they abstain from alcohol). Describe the sampling distribution of $\hat{p}$.

The student gave the following answer:

$$SD(\hat{p}) = \sqrt{(.2)(.8)/85} = .0434$$

Is this answer complete? Explain.

2. Suppose a student claims to have gotten a sample proportion of .38 orange in a sample of 25 candies. Explain how you know that he is mistaken.

3. A student is told that that 73% of all flowers bought for Valentine's Day are bought by men and is asked to identify the parameter. Explain what is wrong with each of the following answers:

a. Men who buy flowers on Valentine's Day

b. All flowers bought on Valentine's Day

c. Whether or not the flowers were bought by a man

d. The proportion who buy flowers on Valentine's Day

4. Suppose that 70% of all Americans take a shower more often than they take a bath. Then suppose we simulate 1000 samples with 80 people in each sample, recording the sample proportion who take a shower more often than a bath for each sample. Finally, suppose that a student is asked to determine the standard deviation of the 1000 simulated sample proportions, and he calculates

$$\sqrt{\frac{(.7)(.3)}{1000}} \approx .0145.$$

Identify the error in this calculation.

Solutions:

1. This student has only discussed one aspect of the distribution. Remember that to describe a distribution you should discuss shape, center, and spread. Calculating the standard deviation only relates to the spread of the distribution. The

distribution means the overall pattern of the outcomes. A more complete answer would indicate that the sampling distribution of the sample proportion $\hat{p}$ is approximately normal with mean .2 and standard deviation .0434:

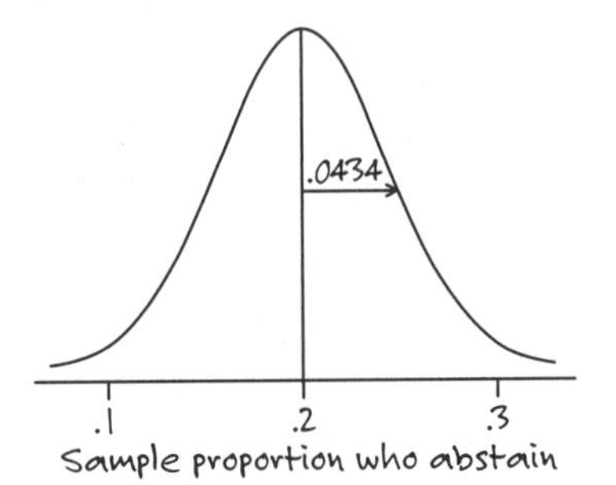

This student did correctly recognize that $\hat{p}$ denotes the sample proportion and then used the population parameter .2 in the standard deviation calculation. Still, it is not clear from this answer that the student understands what this quantity represents.

2. There is no number (integer) that gives .38 when divided by 25. The only possible values for the sample proportion in this case are multiples of .04, such as .36 and .40.

3. The first three choices are automatically wrong because they do not describe a number. Your description of a parameter must always refer to a number. Answer b is a good definition of the population, and answer c is a good description of the underlying variable. Answer d is closer to the parameter in that it is a number, but it is not clear how that proportion is being calculated—the proportion of what? What is the "success"? If the answer had been extended to "the proportion who buy flowers on Valentine's Day that are male," it would have been more accurate. The answer should also clearly indicate that you are talking about the population of all flower buyers so that it is not confused with a sample proportion. Therefore the best answer is "the proportion of the population of all flowers bought for Valentine's Day that were purchased by men."

4. This student has confused the number of samples for the sample size. The sample size is $n = 80$, so the standard deviation of those 1000 sample proportions should be about

$$\sqrt{\frac{(.7)(.3)}{80}} \approx .0512.$$

Topic 17:

SAMPLING DISTRIBUTIONS II: MEANS

This topic paralleled the previous topic but discussed the behavior of the sample mean instead of the sample proportion. The discussion in this topic is relevant when the initial variable of interest is quantitative instead of categorical. You were again led to discover the pattern of these sample means under certain conditions.

Example: You saw in *WS* Activity 15-2 that the weights of newborns follow a normal distribution with mean 3250 grams and standard deviation 550 grams.

a. Suppose you take a random sample of 100 newborns and examine their weights. Sketch a graph of what you expect this distribution of weights to look like. Make sure you label the horizontal axis and indicate the shape, center, and spread of this distribution.

b. Suppose you take 1000 random samples, each of 100 newborns, and examine the mean weight of each sample. Sketch a graph of what you expect this distribution to look like. Make sure you label the horizontal axis and indicate the shape, center, and spread of this distribution.

17

The following graph displays the weights for a population of 72 guinea pigs:

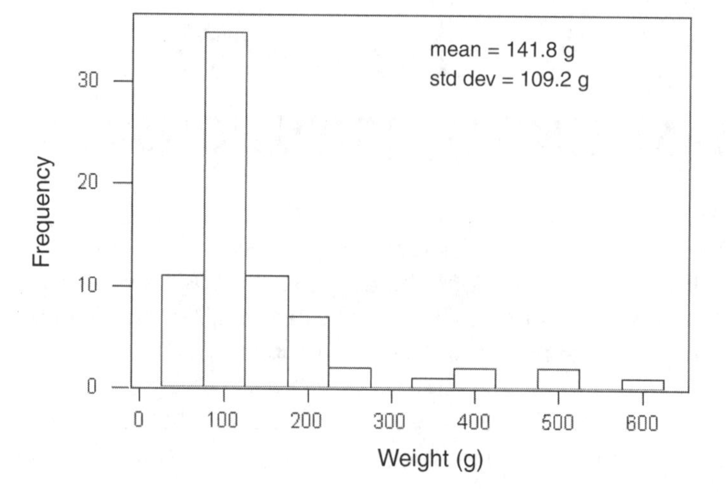

c. Describe this distribution. If you had to replace the label of the vertical axis with the phrase "number of ______," how would you fill in the blank? What symbols can you use to refer to the mean and standard deviation specified in the graph?

d. Suppose you took a random sample of two guinea pigs from this distribution and calculated their average weight and then repeated this process 500 times. Sketch a graph of what you expect this distribution to look like. In particular, how will the shape, center, and spread compare to that of the previous graph? How will you label the horizontal and vertical axes?

Solution:

a. Since the population distribution is normal we would expect the distribution of the sample to look normal also. The mean will be similar to the population mean, and the standard deviation of the sample will be similar to that of the population. That is, we expect $\bar{x} \approx 3250$ grams and $s \approx 550$ grams. The horizontal axis label will be "weight."

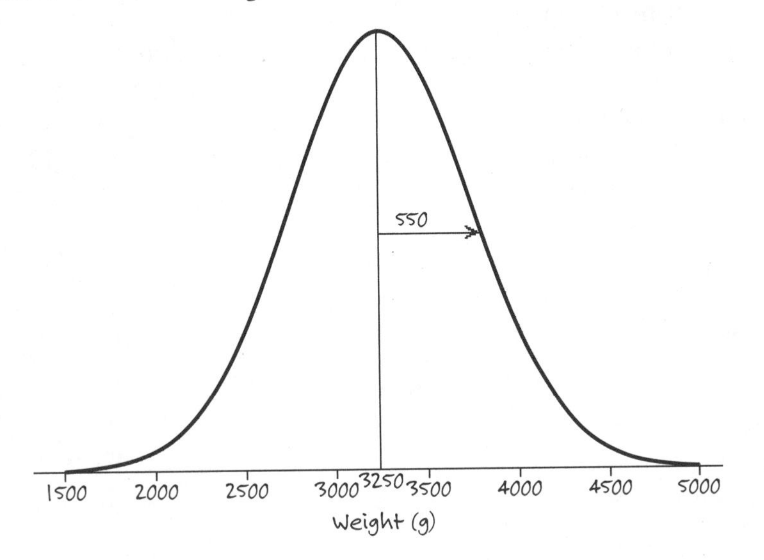

b. Since the population distribution is normal, we would expect the sampling distribution of sample means to be normal also. (In fact, since n is so large, this would be true even if the population had not been normal.) The mean of the sampling distribution will be similar to the population mean, about 3250 grams, but the standard deviation will be smaller, about $550/\sqrt{100} = 55$ grams. The horizontal axis should be labeled "sample mean" or "average weight in sample" or "$\bar{x}$."

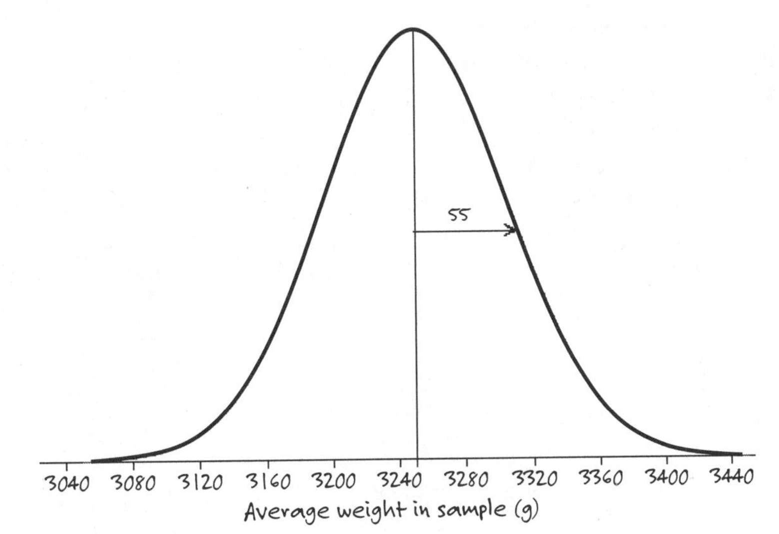

c. This population distribution is skewed to the right with a few outliers corresponding to extra heavy guinea pigs. The mean of the distribution, μ, is 141.8 grams (though the median might better describe a typical weight in this study), and the standard deviation of the distribution, σ, is 109.2 grams (though the interquartile range is a more resistant measure of spread). We could label the vertical axis "number of guinea pigs."

d. Even with a sample size of only two guinea pigs, the sampling distribution will show some differences compared to the population distribution. The mean of the sample means will still be around 141.8, but the standard deviation will be smaller. The standard deviation of the sample means will be about $109.2/\sqrt{2} = 77.2$. The shape will still have a strong skew to the right, but not quite as extreme as in the population. The following is such a graph for 500 different random samples:

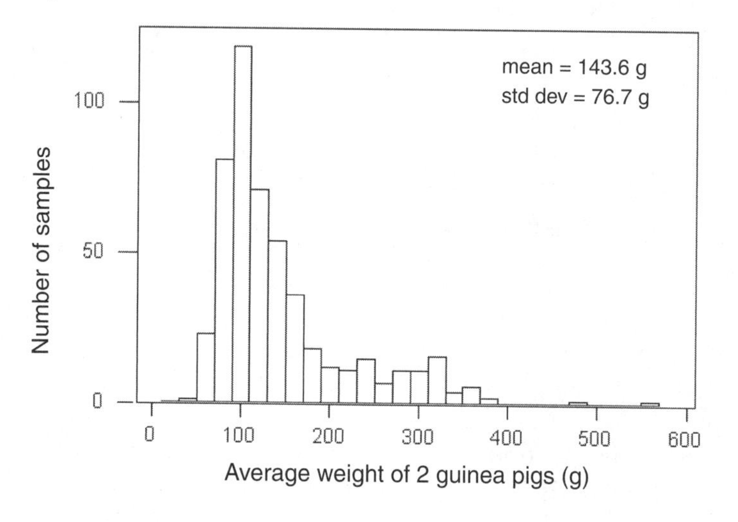

KEY CONCEPTS

- You now know how sample means ($\bar{x}$) can be expected to behave if you take many, many random samples from the same population. In particular, they should cluster symmetrically about the population mean, μ, but with less variability than in the population. The larger you make the sample, the more closely you can expect the sample means to cluster around the population mean. If the sample size is large (e.g., $n \geq 30$) or if the data come from a population that itself has a normal distribution, then the shape of the distribution of the sample means will also be normal.

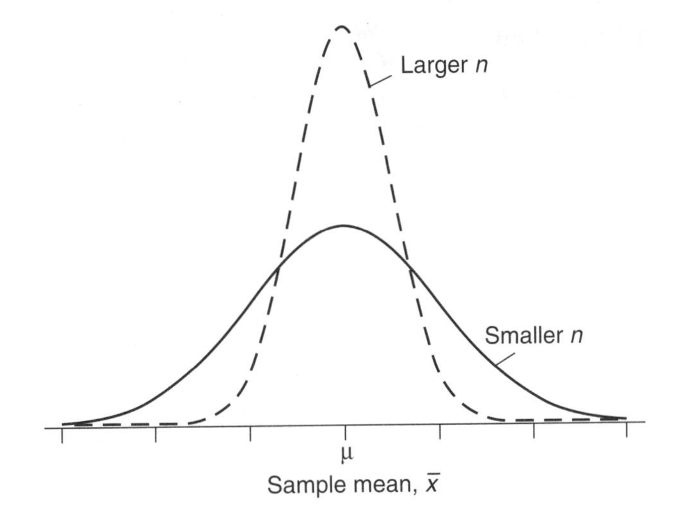

- It is very important for you to keep three distributions in mind:
  - *Population distribution*: the (typically) unobservable distribution. You want to be able to make statements about the population based on what you observe in the sample. This population can consist of categorical data (e.g., yes's and no's, as in Topic 16) or of quantitative data (e.g., weights, as in Topic 17).
  - *Sample distribution*: the actual data that you observe. You have studied numerical and graphical summaries for describing this distribution to others. Roughly speaking, the sample distribution should resemble the population distribution.
  - *Sampling distribution*: the behavior of a sample statistic ($\hat{p}$ or $\bar{x}$) in many, many samples from the same population. You have simulated these distributions. Now you will appeal to the Central Limit Theorem (when it applies) to tell you about this distribution. This information will enable you to make probability statements about obtaining certain values for the sample statistics (e.g., the probability that $\bar{x}$ is between 2.29 and 2.31).
- Again, you also need to be careful in talking about the *number of samples* versus the *number of observations in the sample, n*. The important feature is n; this value will help determine the shape and spread of the sampling distribution. From now on, assume there are always lots and lots of samples. For each distribution in the previous bulleted list, ask yourself how it would be affected if you increased the sample size, the population size, or the number of samples.

COMMON OVERSIGHTS

- Not changing the standard deviation when the sample size differs from 1. This error is a very common oversight students make. The mistake appears to stem from confusion over the sample and the sampling distribution, and not recognizing when the problem requires that you calculate probabilities about the *sample mean* instead of about an individual observation. Always ask yourself: "standard deviation *of what*?" The standard deviation of the individual measurements is σ, but the standard deviation of the sample means is $\sigma/\sqrt{n}$.

FURTHER EXPLORATION

The "Sampling Pennies" Java applet at the following address provides a visual interpretation of the sampling pennies activity: *www.rossmanchance.com/applets/SampleData/SampleData.html*. The top right window displays the distribution of the mint years for a population of 1000 pennies.

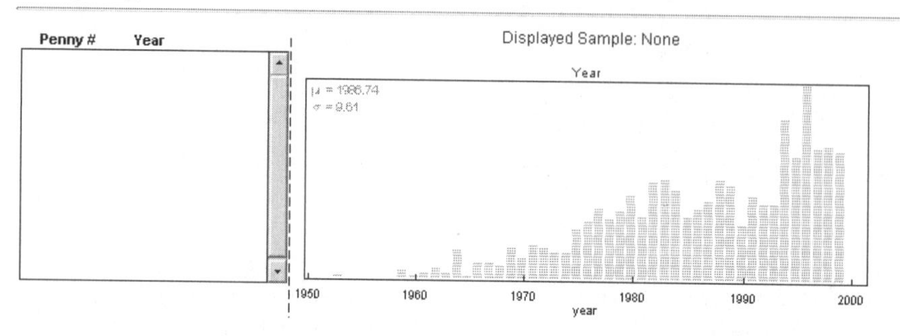

a. Describe this population distribution of years (shape, center, and spread). [*Note:* the applet reports some values in the upper left corner of this window.] Be sure to state your comments in context.

b. Specify a sample size of 5, and click the **DRAW SAMPLES** button. Note the sample is reported in the left table and appears in black atop the population distribution. Report the five years that were randomly selected and the mean of this sample.

c. Click **DRAW SAMPLES** again, and report this second sample mean. What is the average of these two values?

(In part c, you should see that the mean of these two sample means is reported by the red arrow in the lower right graph. The standard deviation of the two values is also reported below the lower dotplot.)

d. Click the **RESET** button. Uncheck the **ANIMATE** button and change the number of samples from 1 to 200. Click the **DRAW SAMPLES** button. Sketch and describe the shape, center, and spread of the sampling distribution of the sample means. How does this sampling distribution compare to the population distribution?

e. Repeat part d for a sample size of 10 and then for a sample size of 50. Sketch and describe both of these distributions. How does each compare to the population distribution? How does each compare to the previous distribution?

f. For a sample size of 50, how does the standard deviation of the sampling distribution compare to $\sigma/\sqrt{n}$?

Solution:

a. The distribution of years is skewed to the left with mean $\mu = 1986.74$ and standard deviation $\sigma = 9.61$ years.

b and **c.** You will probably not get the same mean for both samples. But make sure you understand that you must take the average of the averages here.

d. The distribution is mostly symmetric with perhaps a slight skew to the left. The center should be around the population mean value of 1986.74, and the standard deviation of the sample means should be close to $\sigma/\sqrt{n} = 9.61/\sqrt{5} = 4.30$. Thus, the distribution of the sample means is less skewed and less variable than the distribution of years for the individual pennies in the population.

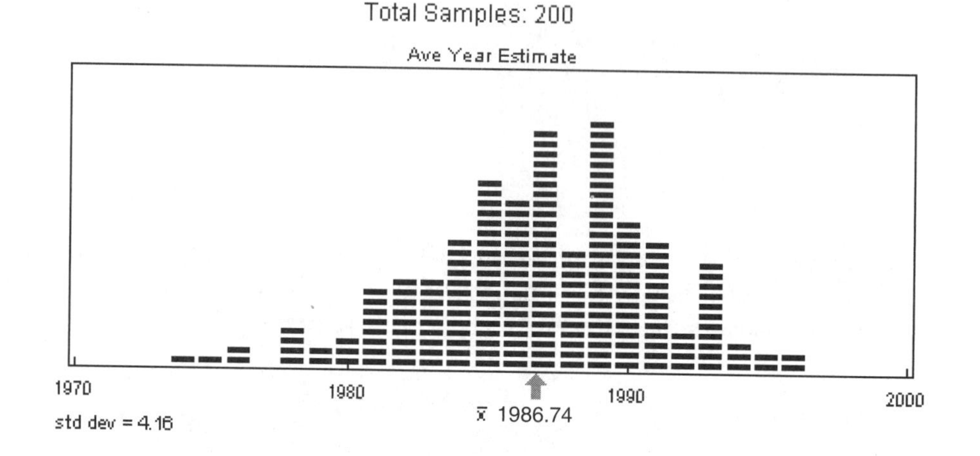

e. As you increase the sample size, you should see that the sampling distribution of the sample means becomes more symmetric and less variable each time. Although the population contains several pennies minted before 1970, it would be very surprising to see a sample mean that small, even for a sample of only 5 pennies.

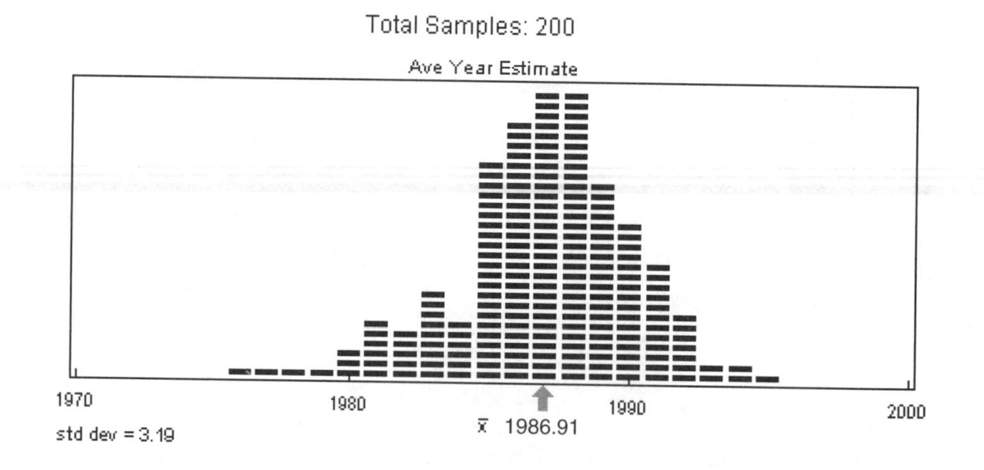

f. With $n = 50$, you should see that the standard deviation of the distribution of sample means is close to $9.61/\sqrt{50} = 1.36$.

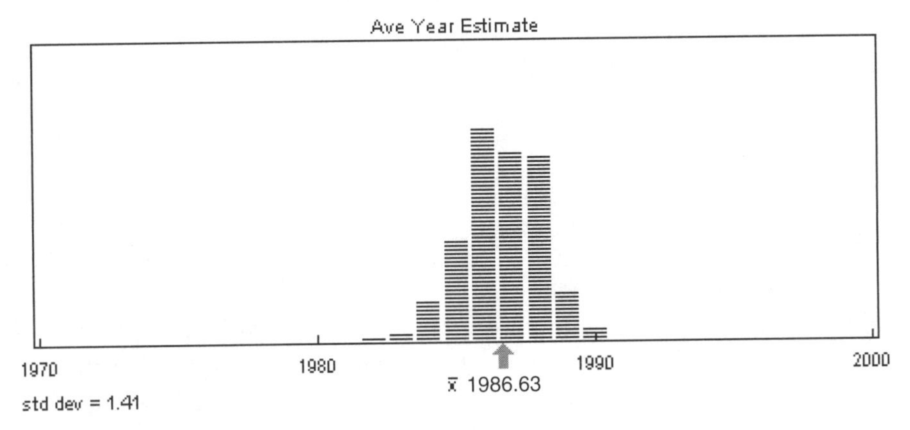

WHAT WENT WRONG?

The following are examples of errors in the analysis of the example problem. In each case, explain what is wrong.

Consider the following question: Suppose that a golfer keeps track of how far she hits the ball when she practices her long shots, and she has found over time that her distances have an average of 180 yards and a standard deviation of 20 yards. Suppose that she starts using a new brand of golf ball and wants to see if her average distance improves. She hits 64 shots with the new ball and calculates her average distance to be 190 yards. Does this provide strong evidence that the mean of her driving distances is greater than 180 with the new ball? Explain.

Identify the flaw in each of the following responses:

1. Yes, because 190 is larger than 180.

2. No, $(190 - 180)/20 = 0.5$, so 190 is less than one standard deviation above the mean of 180. This outcome of 190 would not be a surprising result even if the mean driving distance with the new ball were still 180.

3. You can't tell from this information. Because the driving distance might have a skewed distribution, you can't use the empirical rule or perform any normal probability calculations. You can't determine whether a sample mean of 190 is a surprising result if the population mean were still 180.

Solutions:

1. This response ignores sampling variability completely. It's not good enough to compare 190 to 180, because you need to consider the fact that the sample mean distance will vary from sample to sample. The key issue is how much variability is in the sample mean distances (how far they tend to fall from the population mean based on sampling variability alone) and whether a sample mean distance of 190 yards is surprising if the population mean were 180 yards.
2. This response confuses the standard deviation of the distances with the standard deviation of the sample means. The standard deviation of the sample means is $\sigma/\sqrt{n} = 20/\sqrt{64} = 2.5$ yards. Thus, the observed sample mean of 190 yards is $(190 - 180)/2.5 = 4$ standard deviations above the population mean of 180 yards. Such a result would be extremely unlikely, so the observed sample mean does provide strong evidence that the population mean distance with the new ball is greater than 180.

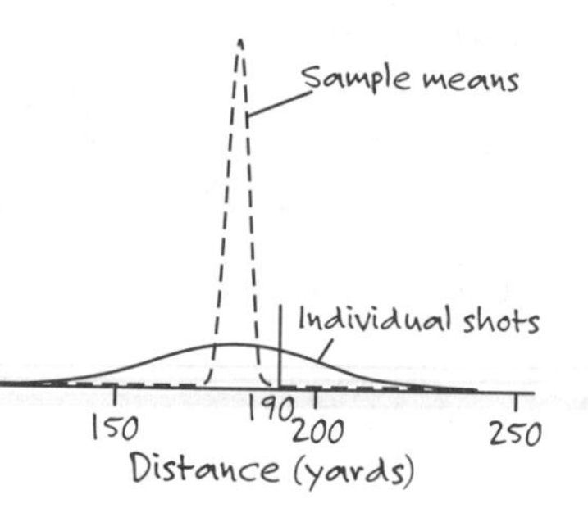

3. This response ignores the fact that with a sample size as large as $n = 64$, the Central Limit Theorem establishes that the sampling distribution of the sample mean will be approximately normal, and so the empirical rule applies, even if the population of distances is not normal.

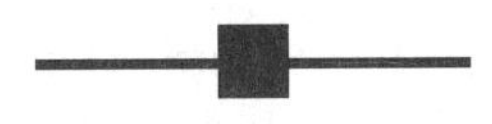

Topic 18:
CENTRAL LIMIT THEOREM

In this topic, you put to use the formulas that you learned in the two preceding topics. When approaching these types of calculations, the first question you need to ask yourself is whether the primary variable is quantitative or categorical. The answer will indicate whether you are working with means or proportions, which will then tell you which set of results to use. The relevant results are summarized in the following table:

Sampling distribution	Categorical variable	Quantitative variable
Sample statistic	$\hat{p}$ = sample proportion	$\bar{x}$ = sample mean
Center (mean)	θ	μ
Standard deviation	$\sqrt{\theta(1-\theta)/n}$	$\sigma/\sqrt{n}$
Shape for large *n*	approximately normal	normal or approximately normal
When is *n* large?	when $n\theta \geq 10$, $n(1-\theta) \geq 10$	when population is normal or $n \geq 30$

Now that you have established when these sampling distributions follow a normal distribution, you can take this a step further and use *WS* Table II to calculate probabilities for obtaining different values of the sample statistic. For example, finding the probability that a sample proportion will exceed .25 or the probability that a sample mean will fall between 2.18 and 2.20 (assuming the sample is selected at random, of course). These probability calculations enable you to make statements like "I would find such a result for the sample proportion surprising," and "I expect to find my sample mean, $\bar{x}$, within two standard deviations of the population mean μ."

In setting up these problems, you should first determine whether the Central Limit Theorem applies. If it does, then specify the mean and standard deviation for the normal distribution that describes the sampling distribution. Sketch and label this distribution and shade the region of interest. Then treat this as a *WS* Topic 15 normal probability calculation by standardizing the observation of interest and using Table II to determine the appropriate probability.

One important concept reinforced by this topic was that sampling variability decreases if a larger sample size is used. This relation enables you to predict whether a probability of interest should increase or decrease with a change in *n*.

Example 1: Recall the scenario about individuals that claim to sleep less than seven hours per night. Suppose you know that 30% of adult Americans will report sleeping less than seven hours each night.

a. If you were to take many random samples from this population, each with 1000 adults, would the Central Limit Theorem apply for the sample proportion? Sketch the sampling distribution predicted by the Central Limit Theorem. Make sure you label the horizontal axis and that you indicate the mean and standard deviation of the distribution.

b. Based on the distribution in part a, calculate the probability of observing a sample proportion of .31 or larger.

c. Would you consider it surprising to obtain a random sample in which at least 31% report that they sleep less than seven hours a night? Explain your reasoning.

d. How would the probability in part b change if you took a sample size of 500 adults instead?

Example 2: Return to the criminal finger lengths scenario. In the earlier example, you considered modeling the population of finger lengths as following a normal distribution with mean 11.5 cm and standard deviation 0.55 cm. One of the probabilities you investigated was randomly selecting a criminal whose index finger was shorter than 12 cm.

a. Suppose you plan to take a random sample of five criminals and calculate the average length of their index fingers. Can you apply the Central Limit Theorem to calculate the probability that this average will be longer than 12 cm? If not, explain why. If so, do so.

b. Will the probability in part a be larger or smaller than the probability that an individual criminal has an index finger length longer than 12 cm?

Example 3: For each of the following situations, first indicate whether it is a question about a sample mean or about a sample proportion and then determine whether the Central Limit Theorem applies. In some cases the answer may not be clear, so be sure to describe your reasoning process.

a. A school principal wants to know whether the average number of absences per student for the first five weeks of the new school year is higher than in the past, which followed a skewed distribution with mean 2 and standard deviation .7. He plans to randomly select 40 students and determine how many times they have been absent in the first five weeks.

b. A student wants to examine what proportion of people use their cell phone between classes, thinking it will be around 20%. She positions herself in front of a large academic building and for the first 50 people who leave the building she notes whether they are talking on the phone or soon use it.

c. A student wants to examine how long he lets his radio alarm play before he hits the snooze button. He has his roommate record the amount of time from when the radio begins playing until he hits the alarm each morning for one month.

Solutions:

1. a. With such a large sample size, 1000(.3) = 300 and 1000(.7) = 700, so both values easily exceed 10 and the theorem applies. Therefore, the distribution of sample proportions will be approximately normal with mean .3 and standard deviation $\sqrt{\theta(1-\theta)/n} = \sqrt{(.3)(.7)/1000} = .0145$.

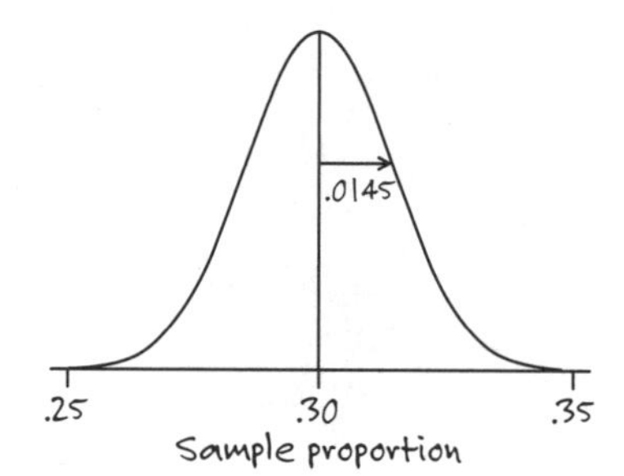

b. Since you are interested in finding a sample proportion of .31 or larger, you will shade your sketch as follows:

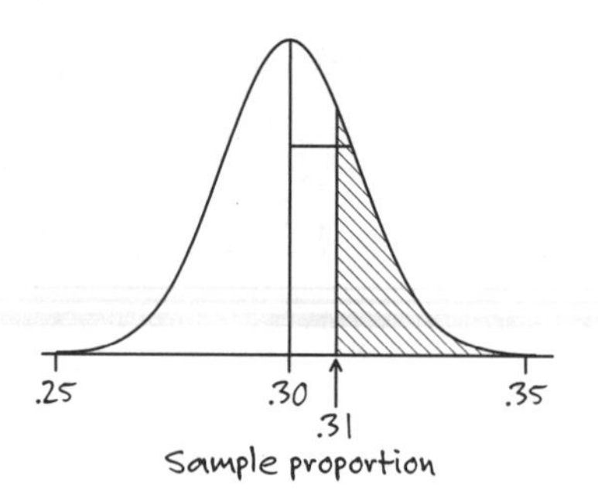

Then standardize this observation:

$$z = \frac{.31 - .3}{.0145} = 0.69$$

So a value for $\hat{p}$ of .31 is .69 standard deviations above the population proportion of .30. Using Table II to find the corresponding probability, you see that the probability below $z = 0.69$ is equal to .7549. To convert this to the probability of interest, you subtract from 1: probability greater than $z = 0.69$ is equal to $1 - .7549 = .2451$.

c. If you took repeated samples from a population with $\theta = .30$, you would observe a $\hat{p} \geq .31$ in 24.51% of the samples. Thus, it is not surprising to get a sample proportion of at least .31 when the population proportion is .30, even with a sample size of 1000.

d. If the sample size had been 500 instead of 1000, then the sampling distribution would have more sampling variability. With the larger standard deviation, the z-score would be smaller, and the probability of obtaining a $\hat{p} \geq .31$ would be larger.

2. a. Since you were told that a random sample was taken, that condition is met. The sample size is small, 5, but since you decided previously that the population appears to follow a normal distribution, you can apply the Central Limit Theorem despite the small sample size. The theorem indicates that the sampling distribution of the sample means is normal with mean equal to 11.5 cm and standard deviation $0.55/\sqrt{5} = 0.246$ cm.

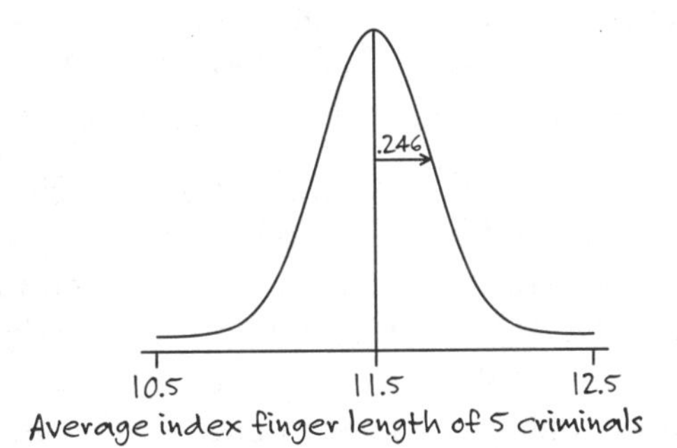

You are interested in an average greater than 12 cm. Standardizing, you get

$$z = \frac{12 - 11.5}{.246} = 2.03,$$

indicating that an average of 12 cm is 2.03 standard deviations above the mean of 11.5 cm.

Using Table II, you find a probability of .9788 below $z = 2.03$, so the probability above $z = 2.03$ is .0212. Thus, if you repeatedly took samples of five criminals, about 2% of the samples would result in an average length greater than 12 cm.

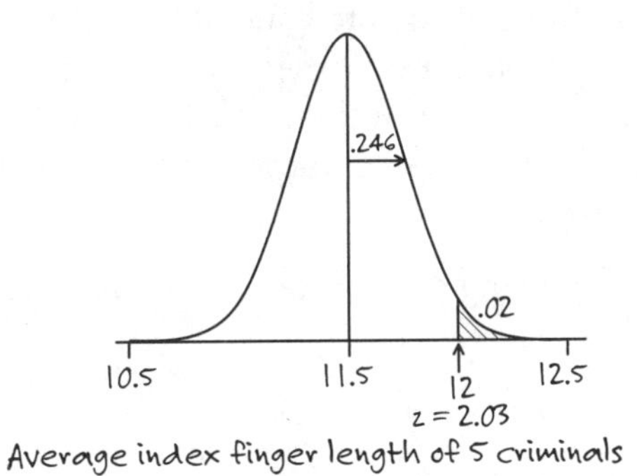

b. The probability that the average finger length will exceed 12 cm will be smaller because the sampling variability for samples of size $n = 5$ is smaller than the sampling variability for samples of size $n = 1$.

3. a. The principal recorded "number of absences," which can be considered a quantitative variable since it makes sense to talk about the average number of absences. So, this is a question about a sample mean. Even though the distribution of the number of absences is highly skewed, this sample size, 40, should be large enough for the Central Limit Theorem to apply. You are also told that a random sample will be taken. Thus, the conditions for the Central Limit Theorem are met.

b. The variable "whether or not they use their cell phone" is categorical. So this is a question about a sample proportion. If you take the population proportion to be $\theta = .20$, then $50(.20) = 10 \geq 10$ and $50(.80) = 40 \geq 10$, so this condition of the theorem is barely met. You should be concerned about whether this sample is representative of all students at the school since the student focused on one building and took only the first 50 students to leave the building. For this reason, you should be cautious in applying the Central Limit Theorem.

c. The variable "amount of time music plays" is quantitative. So this is a question about a sample mean. If the data collection is done every morning for a month, then the sample size is close to 30, so the Central Limit Theorem (barely) applies. If the times themselves follow a normal distribution (you would need to look at a graph of the sample to see whether this appears believable about the population), then there is no doubt that the theorem applies as long as the sample is chosen randomly. However, you might question whether these data will be representative of all the days throughout, say, one year. Maybe there were some unusual or stressful events that month. Still, if he picked the start date at random and didn't try to generalize these results over too long of a time period, you might apply the Central Limit Theorem.

WHAT WENT WRONG?

The following are examples of errors in the analysis of the example problem. In each case, explain what is wrong.

1. A student is given the following problem: In the past, about 10% of people were left-handed. You suspect the percentage is increasing because people are less likely to dissuade a child from being left-handed these days. You observe a random sample of 100 first-graders at recess to see whether they throw with their left or right hand, and you find that 15 use their left hand. What is the probability that you would observe 15 or more first-graders using their left hand if the population proportion was still 10%?

   She presents the following calculations. What has the student done wrong in each case?

a. $z = \dfrac{15 - .10}{.10/\sqrt{100}} = 1490$

probability above ≈ 0

b. $\text{SD} = \sqrt{\dfrac{(.1)(.9)}{100}} = .03$

$z = \dfrac{.15 - .10}{.03/\sqrt{100}} = 16.67$

probability above ≈ 0

c. $z = \dfrac{.10 - .15}{.03} = -1.67$

probability above: $z = -1.67$ is $1 - .0475 = .9525$

2. A student is given the following problem: Weights of eggs follow a normal distribution with mean 56 grams and standard deviation 7 grams. A carton of 12 eggs will be randomly selected from the grocery store display. Each egg will be weighed. What is the probability that the average weight in this sample will be between 50 and 60 grams?

He presents the following analysis. Identify two different mistakes he has made.

> We are told the population of weights is normal and the sample was randomly selected.
>
> $$z = \frac{50 - 56}{7} = -.86$$
>
> probability below = .1949
>
> probability between 50 and 60 = 2(1 − .1949) = 1.61

Solutions:

1 a. The student appears to be confused about whether this is a problem about a mean or about a proportion. The formula used here is the formula for standardizing a sample mean. However, 15 is a *count*, not a sample mean. Thus, the student needs to first convert the 15 to a *proportion* (.15) and then use the standard deviation formula for proportions $\sqrt{.1(.9)/100} = .03$. Then, standardizing, she will get $z = (.15 - .10)/.03 = 1.67$. Per Table II, the probability above this value is .0475. The conversion of z to a probability is valid because the Central Limit Theorem applies here (barely!) since $n\theta = 100(.10) = 10 \geq 10$ and $n(1 - \theta) = 100(.90) = 90 \geq 10$ and the problem stated that the first-graders were randomly selected.

[*Note*: It would be reasonable to compare a count of 15 to 10 but you would need to use a different standard deviation formula, which you have not studied! So always convert the results for a categorical variable into a proportion instead of working with the count.]

b. The student is still trying to use the formula for standardizing a sample mean. The sample size has already been taken into account in the standard deviation calculation, so she should not be dividing by $\sqrt{100}$ again in the denominator.

c. The only mistake here is that she has confused the observation with the mean of the distribution. The numerator of the z-score should always be observation − mean. Since the proportion of left-handers, .15, is larger than the given population proportion, .10, the z-value should be positive. Therefore the probability above should be less than .5.

These types of calculation errors are much less likely to occur if you start with a sketch of the sampling distribution first.

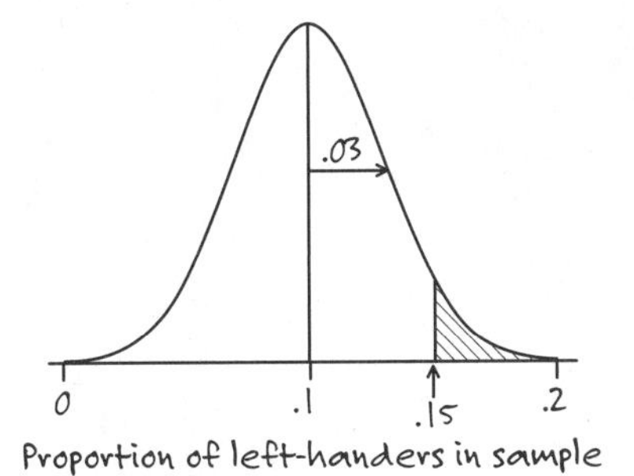

A sketch helps you see that you want to compare .15 to .1 in terms of the .03 standard deviation, and that the resulting probability should be fairly small.

2. The student has assumed that the problem involves an individual egg rather than a sample of 12 eggs. The sample size in this calculation is $n = 12$, not $n = 1$. This affects the calculation of the standard deviation, $\sigma/\sqrt{n} = 7/\sqrt{12} = 2.02$. So when you standardize, you get

$$z = \frac{50 - 56}{2.02} = -2.97$$

probability that the sample mean weight is less than 50 grams = .0015.

To find the probability that the sample mean is between 50 and 60 grams, you also need to find the probability below 60:

$$z = \frac{60 - 56}{2.02} = 1.98$$

probability that the sample mean is less than 60 grams = .9761.

Then you find the probability between 50 and 60 grams by subtracting: .9761 − .0015 = .9746.

You should note that the two "tail" probabilities (below 50 and above 60) are not equal here since the mean is not exactly halfway between them. You should have recognized immediately that there was a problem with the calculation when the probability came out to be greater than 1!

Again, a picture is very helpful here:

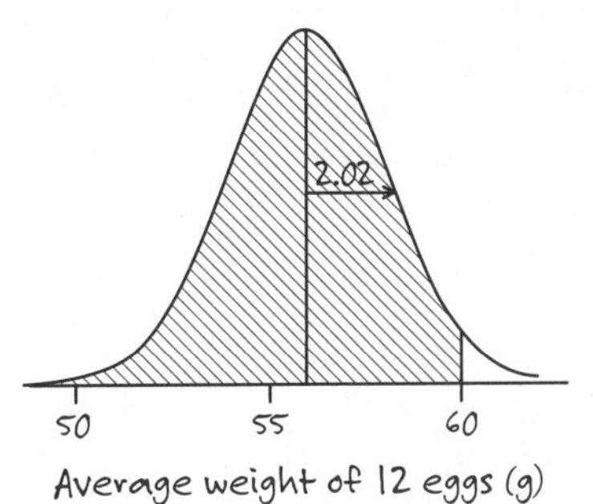

Unit V:

Inference from Data: Principles

Topic 19:
CONFIDENCE INTERVALS I: PROPORTIONS

Before you begin Unit V, you might want to review the following:

- The distinction between *population* and *sample*. Remember that the population is the entire group you are interested in, whereas the sample is a subset of that group from which you actually collect data.
- The importance of taking a *random* sample. Taking a random sample from the population gives every member of the population the same chance of being selected and enables you to predict what the pattern of variation in sample results will look like from sample to sample.
- The difference between a *parameter* and a *statistic*. A parameter is a number (often a mean or a proportion) that describes the population, whereas a statistic is the corresponding number calculated from sample data. You can calculate the value of a sample statistic, but you usually do not know the value of a population parameter. In fact, the whole point of taking a sample and calculating a statistic is often to estimate the population parameter or make a decision about its value.

Those are the key terms. Even more fundamental, but particularly to this topic, is to recall yet again the idea of a variable and its type. In this unit you worked with both binary categorical variables and with quantitative variables. As you review these ideas, you should start by identifying which type of variable you are working with.

The Gallup Organization (together with CNN and *USA Today*) conducts a poll every few months to determine the U.S. president's approval rating. They typically survey 1000 randomly selected adults, aged 18 or older, and ask questions such as whether the respondent approves or disapproves of the way the president is handling his job. Based on the sample results, they make claims as to the percentage of Americans who approve. How are they able to make conclusions about all American adults, predicated upon just one sample of roughly 1000 respondents?

In this topic, you formalized the procedure for calculating a confidence interval for one population proportion. It is important to keep the primary goal in mind: determining plausible values for the population proportion based on your observation for the sample proportion.

Example: A CNN/*USA Today*/Gallup poll conducted July 25–27, 2003, with a randomly selected sample of 1003 adults, found that 58% of Americans "approved of the way George Bush was handling his job as president." Bush's ratings had been in the low 60's in the preceding months, and the historical average presidential approval rating (dating back to the 1940s) is 56%. Summarize these results and estimate the proportion of the population of all adult Americans who approved of the job Bush was doing at this time.

Solution:

In this problem, you are estimating the population parameter based on a sample result. You don't expect the population result to be 58% exactly, but you believe it will be close. Therefore, you will construct a confidence interval that specifies a range of plausible values for the population parameter. In solving these problems, the first step is to identify the observational units and the primary variable(s) involved. This process will help you determine the appropriate analysis.

The data and parameter

The observational units are the adults (age 18 or older), and the variable recorded about each adult is whether or not they approve. This is a binary variable, so you know you will be working with proportions. Let $\hat{p}$ refer to the sample proportion and θ the population proportion. The sample is the 1003 adults interviewed, and, since they were randomly selected from all American adults, the population is all American adults. The poll yielded $\hat{p} = .58$, the proportion in the sample who approve, and you want to estimate θ, the proportion of the population who approve.

It is important to translate everything from the problem statement into the appropriate symbols. Defining the variable first helps you determine whether the problem deals with means or proportions. It is also important to be clear about what is known (the value of $\hat{p}$) and what is unknown (θ, which you can only define in words).

Numerical and Graphical Summaries

At this stage it is helpful to describe the sample numerically and graphically. For a single categorical variable, the appropriate graph is a bar graph.

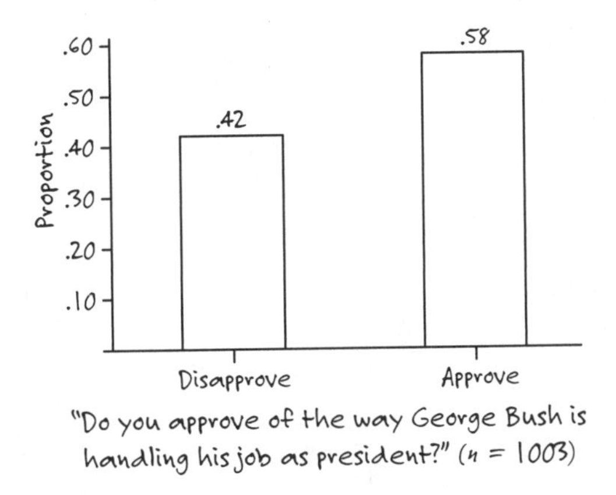

From this graph, you see that the people in the *sample* are slightly more likely to have approved than disapproved in this poll.

Inference

Can you say more than just how the people in the sample felt? You want to make some statement about the proportion in the *population* who approved at this time. Before proceeding with any calculations, you must check the technical conditions to see whether your inference method is valid here.

Technical condition. With one binary variable, you must have a simple random sample from the population of interest and a large enough sample size so that the sampling distribution of the sample proportion is approximately normal. The latter condition will be considered met if

$$n\hat{p} \geq 10 \text{ and } n(1-\hat{p}) \geq 10.$$

Both inequalities must be met. In this case, $n = 1003$ and $\hat{p} = .58$, and $.58(1003) = 582$ and $.42(1003) = 421$ both exceed 10. You might notice that these two values are the number of successes and the number of failures in the sample. The Gallup website states "These results are based on telephone interviews with a randomly selected national sample of 1003 adults." So you can consider the random selection condition met as well.

Selecting the appropriate formula. Since the technical conditions are met, you can calculate a confidence interval for θ using $\hat{p} \pm z^* \sqrt{\hat{p}(1-\hat{p})/n}$.

Using the formula. Since a confidence level was not specified, we will use 95%. For 95% confidence, Table II or the table at the end of Activity 19-2 tells you that $z^* = 1.96$.

$$\text{Lower endpoint: } .58 - 1.96\sqrt{(.58)(1-.58)/1003} = .58 - .0305 = .5495$$
$$\text{Upper endpoint: } .58 + .196\sqrt{(.58)(1-.58)/1003} = .58 + .0305 = .6105$$

Notice that you entered the sample *proportion* (.58) into the formula, not the sample *percentage* (58).

Summarizing the results. You are 95% confident that θ, the proportion of the population who approved of Bush at this time, is in the interval (.5495, .6105), or that between roughly 55% and 61% of the population approved. Thus, you conclude that a small majority of adult Americans approved of Bush at this time. You also notice that the interval includes .56, indicating that Bush's rating was not significantly higher than the historical average rating.

Interpreting confidence (not always required). If you are asked what is meant by the phrase "95% confidence," you would say that if someone repeatedly took random samples of 1003 adult Americans and calculated a 95% confidence interval for each sample, then in the long run, roughly 95% of those intervals would succeed in capturing the actual value of the population proportion.

Notice that this interpretation doesn't even mention the particular interval values that you calculated. Interpreting the 95% confidence interval applies to the general method, not the one specific interval you calculated.

KEY CONCEPTS

- The following list summarizes the steps and calculations for a confidence interval for θ.

  1. Define the parameter θ in words and the population of interest.
  2. Check the technical conditions for the validity of the procedure:
     - Simple random sampling
     - $n\hat{p} \geq 10$ and $n(1 - \hat{p}) \geq 10$
  3. State the confidence interval formula and perform the calculations:

     $$\hat{p} \pm z^* \sqrt{\hat{p}(1 - \hat{p})/n}$$

  4. Provide a statement about the interval in the context of the problem.

- The general form of a confidence interval is estimate $\pm$ margin of error, where margin of error = (critical value) $\times$ (standard error of estimate).
  - Note that the margin of error is always equal to half the length (upper endpoint − lower endpoint) of the interval.
  - For a single proportion, the margin of error is equal to $z^* \sqrt{\hat{p}(1 - \hat{p})/n}$.
- The margin of error relates only to the amount of sampling variability and the confidence level.
  - It does not tell you about nonrandom sampling errors (bias) or nonsampling errors (e.g., errors caused by the subject's lying or exaggerating or by the interviewer's effect on the subject's response).
- Several factors affect the width of a confidence interval.
  - Larger samples produce smaller widths (narrower intervals) since the standard error of the estimate ($\sqrt{\hat{p}(1 - \hat{p})/n}$) decreases as n increases.

- Higher confidence levels lead to larger widths (wider intervals) since the critical value, z^*, increases.
- Population size does not affect the width of a confidence interval.

- Ideally, you will have a narrow interval and a high confidence level (precise and with a strong belief that it contains the parameter), but there is a trade-off. If you only increase the confidence level, you will have a wider interval. You can always take a larger sample to produce a narrower interval without decreasing the confidence level, but this usually requires more time, money, and effort.

CALCULATION HINTS

- Show the numbers plugged into the formula. This makes it easier to break the calculation into several steps, decreasing your calculation errors.
  - Remember to use the *proportion*, not the percentage or the count. In particular, $1 - \hat{p}$ (the proportion of failures in the sample) needs to be a number between 0 and 1 for you to be able to take the square root.
  - Remember to take the square root *after* you have worked out the $\hat{p}(1 - \hat{p})/n$ part.
  - Don't forget to multiply the square root part by z^*.
  - Remember to multiply and divide before you add or subtract.
- Calculate the margin of error first. Then add and subtract this amount from the statistic.
- Carry a few extra digits, say four or five decimal places, and then round to two or three places at the very end of the calculations.
- When you are done calculating the interval, make sure the values are reasonable. For example, since you are estimating a proportion, are the numbers between 0 and 1? Is the sample proportion (.58 in the example) in the middle of the interval? In the example, the sample size was fairly large, so the interval was pretty narrow.

COMMON OVERSIGHTS

- Not realizing that you have calculated an interval.
  - Remember, the interval specifies a range of values, all of the values between the two endpoints you calculated. Any number in the interval is plausible as a value for the population parameter. Numbers outside this range, even .56 in the example, are not considered plausible values for θ at that confidence level.
- Not being clear that the confidence interval relates to the population parameter.
  - The interval does not estimate the value of the sample proportion. You know the value of $\hat{p}$, and you know that it is *always* inside the interval you calculate. In fact, it is always the midpoint of the interval.
- Misinterpreting what is meant by the term "confidence."
  - You can't say that you are 95% confident that the population is between the two endpoint values of the interval. The population consists of yes's and no's, so you are not capturing population values with the interval. The interval claims to capture the population proportion, the numerical characteristic of the population of interest.
  - Pay very close attention to the statements at the end of *WS* Activity 19-5 on how to interpret confidence (and how not to).

- Failing to check the technical conditions.
  - The interval formula is only valid if you have simple random sampling and if n is large. If either of these conditions is not satisfied, you can't safely interpret the confidence interval calculation.
- Not writing a one-sentence interpretation of your calculations in the context of the study.

WHAT WENT WRONG?

The following are examples of errors in the analysis of the example problem. In each case, explain what is wrong.

1. We are 95% confident that between 55% and 61% of the 1003 adults surveyed approved of Bush.

2. The confidence interval for θ is $58 \pm 1.96\sqrt{58(1-58)/1003}$.

3. If you repeatedly sampled 1003 adult Americans, roughly 95% of these sample proportions would be between .5495 and .6105.

4. Bush was doing better than the historical average because .56 is close to the lower endpoint.

5. The confidence interval for θ is $.58 \pm 1.96\sqrt{.58(1 - .58)/1003} = (-.0215, .0396)$.

6. The confidence interval for θ is $.58 \pm 1.96\sqrt{.58(1 - .58)/1003} = (.5795, .5804)$.

7. The confidence interval for θ is $.58 \pm 1.96\sqrt{.58(1 - .58)/1003} = .58 \pm .0156$.

8. The confidence interval for θ is $1.96\sqrt{.58(1 - .58)/1003} = .0306$.

Solutions:

1. This statement refers to the sample proportion instead of the population proportion, the value the 95% confidence interval is designed to capture.
2. This solution is using the percentage, 58%, instead of the sample proportion, $\hat{p} = .58$. This calculation would be impossible to work out because it would have a negative value under the square root.
3. This statement is incorrect since it refers to the endpoints of one specific interval. If you repeatedly sampled 1003 adult Americans and constructed a confidence interval based on each sample, these intervals would vary slightly from sample to sample. The confidence level is not making a statement about the endpoints of the intervals but about what percentage of the intervals you believe will capture the population proportion. "Confidence" refers to the method of determining the intervals, not about any one particular interval.

4. The answer recognizes that .56 is barely inside the interval. In this case, you are 95% confident that θ is somewhere in this interval. Even though the lower endpoint is not much below .56, it is below .56. So the conclusion that follows from this interval is that .56 is a plausible value for θ (with 95% confidence), and we can't conclude θ is above .56.
5. This calculation did not follow the order of operations correctly and added and subtracted 1.96 from .58 before multiplying. You should always work out the part to the right of the $\pm$ first and then work with the sample proportion. You should also be able to recognize that this interval could not have been calculated correctly since the sample proportion, .58, is not in the center.
6. This calculation did not take the square root of $.58(1 - .58)/1003$. You might have spotted this since the confidence interval was extremely narrow. You should expect a sample size of more than 1000 to lead to a small interval, but not quite this small!
7. The square root was calculated correctly, but the calculation failed to include the 1.96 multiplier.
8. This student has not realized that an interval is a range of values. Only the margin of error has been calculated here.

FURTHER EXPLORATION

Open the "Simulating Confidence Intervals" Java applet at *www.rossmanchance.com/applets/Confsim/Confsim.html.*

The confidence interval method you use in this text is called the *Wald* interval, though the applet also allows for exploration of more advanced procedures.

Notice that the population parameter is set to $\theta = .5$, the sample size is set to $n = 100$, and the confidence level is set to 95%.

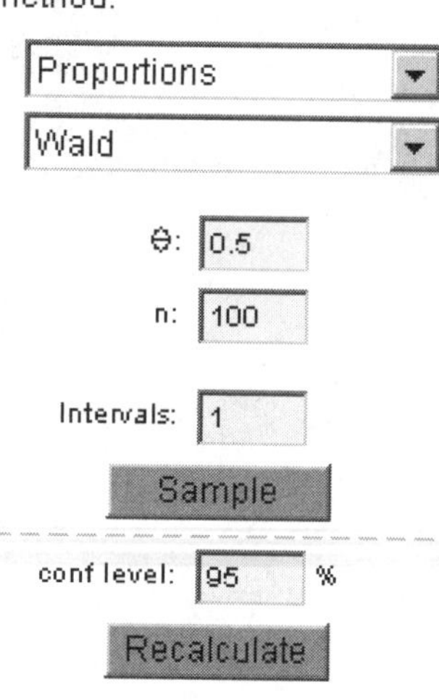

a. Click the **SAMPLE** button. The applet selects a random sample of 100 observations from this population, determines the sample proportion $\hat{p}$, and computes a 95% confidence interval. Click on the line segment representing this interval so the applet reports $\hat{p}$ and the endpoints. Verify the calculation of these endpoints.

b. Change the number of intervals from 1 to 300, and click the **SAMPLE** button. The applet colors an interval green if it succeeds in capturing $\theta = .5$ and red if it fails to capture θ. The applet also reports the percentage of the 300 intervals that capture θ. What percentage of your 300 intervals capture θ?

c. Change the confidence level from 95% to 90% and click **RECALCULATE.** Now what percentage of intervals succeed in capturing θ? What characteristic about the intervals themselves changes?

d. Change the sample size from $n = 100$ to $n = 200$, and click **SAMPLE.** What happens to the widths of the intervals? Does the percentage of intervals that succeed in capturing θ change much?

Solution:

In the exploration you should have seen that if you repeatedly sample from the same population, the confidence level indicates the long-run percentage of intervals that succeed in capturing the parameter. In part a, about 95% of your 300 intervals should have contained θ. Your percentage may not have been 95% exactly because you took only 300 intervals, and not all possible intervals. In part b you confirmed that if you decrease the confidence level, then, on average, a smaller percentage of the random intervals will capture the parameter. The widths of the intervals are smaller in that case, so they capture θ less often. Another way to decrease the width of the confidence intervals is to increase the sample size, as you confirmed in part c.

Some things to note from this exploration:

- The value of the *population* parameter does not change from sample to sample. It's always represented in this applet by the vertical line.
- What does change from sample to sample is the value of the *sample* proportion. And since the confidence intervals are calculated based on the sample proportion, the resulting interval also changes from sample to sample.
- The confidence level determines how often the procedure succeeds (i.e., what percentage of the intervals are green in the long run).

- The sample size affects the width of the intervals but not the percentage that succeed.

A final thing to note from this simulation is that it's meant to help you understand that "confidence" refers to what happens when you take many samples. In practice you just get one sample. You don't know what the actual parameter value is, but you can be confident that your interval succeeds in estimating the parameter since most intervals do.

Topic 20:

CONFIDENCE INTERVALS II: MEANS

This topic is very similar to *WS* Topic 19 but concerned the population *mean* instead of the population *proportion*. So one of the first questions you should ask yourself is whether the variable of interest is categorical (as in Topic 19) or quantitative.

Example: A sample of 36 statistics students reported on the amount of sleep they had the night before (in hours, to the nearest quarter hour). The data were

```
3.00  4.00  5.00  5.00  5.00  5.00  4.25  4.75  4.75  4.75  5.25
5.75  5.75  5.75  6.00  6.25  6.25  6.50  6.50  6.50  6.50  6.50
6.75  7.00  7.00  7.00  7.25  7.25  7.25  7.50  7.50  7.50  8.00
8.25  8.25  9.00
```

Construct and interpret a 90% confidence interval for the mean amount of sleep for statistics students at this school.

Solution:

As in the example in *TK* Topic 19, your goal is to make a statement about the population (all statistics students at this school) based on what was observed in the sample (the 36 who responded).

The data and parameter

The observational units are statistics students, and the variable recorded about each student is the amount of sleep they got the night before. This is a quantitative variable, so you know you will be working with means. Let $\bar{x}$ refer to the sample mean that you observed, and define μ to be the average amount of sleep for all statistics students (the parameter) at this school, which is unknown to you.

Numerical and graphical summaries

Since there is one quantitative variable, the appropriate graph could be a stemplot, histogram, or boxplot. You should also report the sample mean and the sample standard deviation.

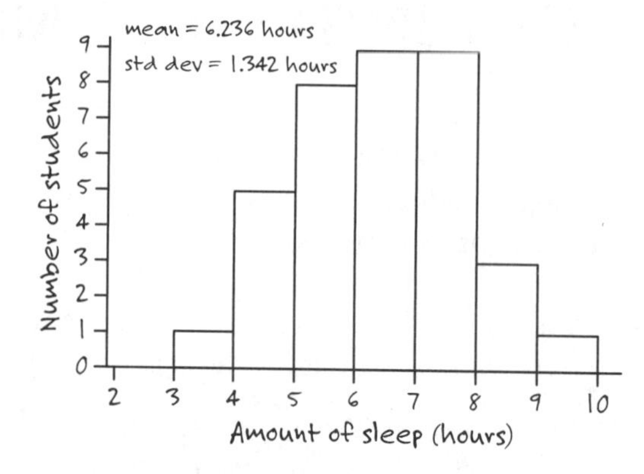

From these results, you can see that the 36 students in the sample averaged 6.236 hours of sleep, with standard deviation 1.342 hours. The distribution is roughly symmetric, ranging from 3 to 9 hours.

Inference

Before using the sample results to estimate the mean amount of sleep in the population, you will check the technical conditions of the interval procedure.

Technical conditions. With one quantitative variable, you need a simple random sample from the population of interest, with either a normal population distribution or a large sample size. The typical cut-off for the sample size is $n \geq 30$. This sample size is 36 students, which is greater than 30 (barely), and the distribution of sleeping times for the sample is reasonably symmetric, so we will consider that condition met. However, no information was given about how the sample was selected. Were all statistics students equally likely to be selected? Was this just one class? Is there any reason why these students might have obtained more or less sleep than the rest of the population? For example, if this was an 8 A.M. class, there could be a problem. Proceed with the calculations, but be very cautious about generalizing this sample to all statistics students at this school until you can obtain more information about how the sample was selected.

If the sample size had been smaller, you would not have been able to appeal to the $n \geq 30$ criterion. Instead, you would have had to rely on the graph of the distribution of sample observations. If you don't see extreme skewness or severe outliers in the sample and it is reasonable that the variable follows a normal distribution in the population (e.g., as you would expect with biological characteristics), then you can proceed even with a small sample.

Selecting the appropriate formula. Since there is one quantitative variable and you are interested in estimating the population mean, you will use a one-sample t-interval:

$$\bar{x} \pm t^* \left(\frac{s}{\sqrt{n}} \right)$$

Using the formula. With a 90% confidence level and a sample size of $n = 36$, the degrees of freedom are $36 - 1 = 35$. So the t^*-value ≈ 1.69 (from *WS* Table III).

Lower endpoint: $6.236 - 1.69\,(1.342/\sqrt{36}) = 6.236 - .3780 = 5.858$
Upper endpoint: $6.236 + 1.69\,(1.342/\sqrt{36}) = 6.236 + .3780 = 6.614$

Summarizing the results. You are 90% confident that μ, the average amount of sleep in the population of statistics students at this school, is between 5.858 hours and 6.614 hours. However, you don't know how this sample is selected, and there could be reasons it is not representative of the entire population. It is very risky to make statements about the entire population based on this sample.

KEY CONCEPTS

- The following list summarizes the steps and calculations for a confidence interval for μ:
  1. Define the parameter μ and the population of interest in words.
  2. Check the technical conditions for the validity of the procedure:
     - Simple random sampling and either a normal population or $n \geq 30$
  3. State the confidence interval formula and perform the calculations:

     $$\bar{x} \pm t^*\left(\frac{s}{\sqrt{n}}\right)$$

  4. Provide a statement about the interval in the context of the problem.
- You should realize that the overall idea of a confidence interval is the same, whether you are talking about the population mean or the population proportion.
  - The interpretations of margin of error and confidence are the same.
  - Don't forget to check the technical conditions before calculating the interval.
    - If you don't have a large sample size, then you need to look at the data and determine whether the sample distribution is reasonably symmetric. Graph the sample data and comment on the distribution.
  - Sample size and confidence level affect the width (margin of error $= t^*\, s/\sqrt{n}$) of the confidence interval with sample means in the same way as with sample proportions.
    - In addition, the variability in the sample affects the width of the confidence interval for μ. If the data are less variable (smaller standard deviation), then your estimate will be more precise (smaller margin of error, narrower interval).

CALCULATION HINTS

- See Calculation Hints for *TK* Topic 19.
- A confidence interval for the population mean will *always* use a critical value from the t-distribution in this text.

- If the degrees of freedom you are looking for are not in the table, the convention is to round down to the nearest value that does appear in the table. This makes the interval slightly larger than it needs to be but guarantees that you are at least as confident as you claim to be.
- When you are done calculating the interval, make sure it is reasonable for the context. For example, the sample mean should be in the middle of the interval.

COMMON OVERSIGHTS

- Using the z^* critical value rather than the t^*-value.
- Forgetting to divide by $\sqrt{n}$ in calculating the margin of error.
- Not remembering to follow up your calculation with a one-sentence summary of your conclusion in the context of the problem.
- Using the term "confidence interval" when you mean "confidence level."
- Checking the wrong technical conditions or misstating the conditions; for example:
  - Confusing the conditions for a categorical variable with the conditions for a quantitative variable.
  - Thinking that the sample size always has to be at least 30.
  - Thinking that the sample data have to look normally distributed (only one of these last two situations is required).

WHAT WENT WRONG?

The following are examples of errors in the analysis of the example problem. In each case, explain what is wrong.

1. The confidence interval for μ is $6.236 \pm 1.645(1.342/\sqrt{36})$.

2. The confidence interval for μ is $6.236/36 \pm 1.645\sqrt{(6.236/36)(1 - 6.236/36)/36}$.

3. $t^* = 1.306$

4. We know that $n \geq 30$, so we can assume the population follows a normal distribution.

5. The confidence interval for μ is $6.236 \pm 1.69\ (1.342)$.

6. Of the students in the sample, 90% slept between 5.858 and 6.614 hours.

7. Of the students in the population, 90% slept between 5.858 and 6.614 hours.

Solutions:

1. This is wrong because it uses the z^* critical value for 90% confidence rather than the t^*-value. Whenever you are producing a confidence interval for a mean, use t^*.
2. This is wrong because it attempts to use the confidence interval formula for a proportion rather than the confidence interval formula for a mean.
3. This t^*-value was looked up in Table III incorrectly. It is the value for 80% confidence (with the correct 35 degrees of freedom) rather than 90% confidence. If the confidence level is 90% then the upper tail percentage is 10%/2 = 5%, not 10%.
4. The $n \geq 30$ statement is correct, but you can't conclude from this that the population has a normal distribution. It's certainly possible that the population data are skewed or bimodal or generally have a nonnormal shape. However, the $n \geq 30$ condition does tell us that even if the population is not normal, the distribution of sample means from the population will be approximately normal (according to the Central Limit Theorem), so you can go ahead and proceed with the t-interval calculation.
5. This student has forgotten to divide by the square root of the sample size (36). When you start with $\bar{x}$, then you have to use $s/\sqrt{n}$ as the standard deviation, not just s.
6. This conclusion is incorrect because it does not take into account that the interval estimates the mean sleeping time in the population. In other words, the interval estimates an average value, not individual students' values. If you count how many of the sample sleeping times fall within the interval, you will find a lot fewer than 90% here (7/36, or 19.44% of the sample are in the interval).
7. Even though this conclusion mentions the population and not the sample, it is still incorrect because it refers to individual values rather than the population mean sleeping time.

FURTHER EXPLORATION

Open the "Simulating Confidence Intervals" Java applet at *www.rossmanchance.com/applets/Confsim/Confsim.html*. Change the method from proportions to means and from z with sigma to t. Suppose the population has mean μ = 0 and standard deviation σ = 5. Set the applet to these values. Keep the sample size set to n = 100. Specify that you want to randomly select 300 intervals from this population. Click **SAMPLE**.

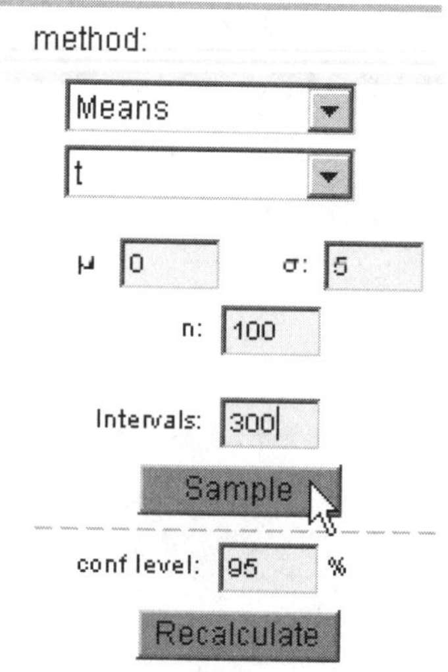

a. What percentage of intervals succeed in capturing μ = 0? Is this what you expected?

b. Change the confidence level from 95% to 90% and click **RECALCULATE**. Now what percentage of intervals succeed in capturing $\mu = 0$? Is this what you expected? How did the width of the intervals change in general?

c. Change the sample size from $n = 100$ to $n = 50$ and click **SAMPLE**. Now what percentage of intervals succeed in capturing $\mu = 0$? Is this what you expected? How did the width of the intervals change in general?

d. Change the population standard deviation from $\sigma = 5$ to $\sigma = 10$. Now what percentage of intervals succeed in capturing $\mu = 0$? Is this what you expected? How did the width of the intervals change in general?

Solution:

a. This percentage should be around 95%, the confidence level.

b. The percentage should decrease to around 90%, because that is the new confidence level. The width of the intervals will be smaller.

c. The percentage should still be similar to the confidence level, but the width of the intervals should be larger.

d. The percentage should still be similar to the confidence level, but the width of the intervals should be larger.

Some things to note from this exploration:

- The confidence level determines the long-run percentage of intervals that succeed in capturing the parameter.
- Decreasing the confidence level again decreases the width of the interval.

- Increasing the sample size decreases the width of the interval but doesn't change the confidence level (the long-run percentage of intervals that you expect to capture the population parameter).
- Increasing the variability in the population increases the width of the interval but doesn't change the confidence level.

You will use the methods presented in Topics 19 and 20 if the goal is to estimate the population parameter based on a sample result. You will recognize these problems by statements such as, "Estimate the proportion . . . ," or, "Determine plausible values for the population mean . . ." You will then have to decide whether the variable of interest is binary (as in Topic 19) or quantitative (as in Topic 20). This determination will help you decide whether you are working with a proportion or an average, even if these words are not used explicitly.

Remember to check the technical conditions before you start any calculations, and that the conditions that need to be checked depend on whether you are working with quantitative or categorical data. For example, in the Gallup poll example about George Bush it would not have been appropriate to only say $n \geq 30$ without also looking at the value of $\hat{p}$. You should clearly state which interval formula you are using and show your work by writing the formula with the numbers plugged in. When you are done with the calculation, include a one- or two-sentence summary in context of what the interval tells you.

Topic 21:

TESTS OF SIGNIFICANCE I: PROPORTIONS

This topic began a discussion of how you can make decisions about a population parameter based on what you observe in the sample. Tests of significance also fall under the heading of "statistical inference," but instead of specifying a range of plausible values for the parameter, as a confidence interval does, tests of significance assess the evidence provided by the sample data about one specific conjectured value for the population parameter.

The hardest part of tests of significance is probably getting your brain wrapped around the main idea.

Example 1: Suppose Kara rolls a pair of dice and looks at the sum of the numbers on the two faces. In her first five rolls, she obtains the following outcomes: 7, 7, 11, 7, 11.

a. What was your first reaction to these results, and why did you react this way?

b. Were you surprised by these outcomes? Explain.

c. It's important to realize that such outcomes are only surprising if you began with what basic assumption about the pair of dice?

Solution:

You probably have some intuition about the behavior of *fair dice*, where all 36 possible outcomes are equally likely to occur. For example, a sum of 7 is the most common outcome (probability 1/6). However, it would be surprising to obtain sums of only 7 and 11 in the first five rolls. In fact, the probability of rolling only 7's and 11's in the first five rolls of fair dice is $(8/36)^5 = .0054$. *If these were fair dice*, Kara was very lucky indeed!

However, you were not told these dice were fair. It is possible that they were not. So now you have to decide between two possible conclusions:

- These dice are fair, and Kara obtained a set of outcomes that happens about 0.5% of the time with fair dice.
- They are not fair dice.

Suppose Kara rolled the dice five more times and got the following sums: 11, 7, 7, 7, 7. Again, you have to decide whether the dice are fair and Kara got even luckier, or the dice are not fair.

Clearly, after ten rolls, there is much more evidence against these dice being fair than after five rolls. We don't know for sure that they are not fair, but the evidence is definitely stacking up against their being fair. In fact, you may now feel there is enough information to feel very comfortable in deciding that the dice were not fair.

You used the following steps in this scenario:

1. State an initial conjecture or hypothesis about the process ("the dice are fair").
2. Gather some data (the five or ten rolls).
3. Ask how likely the sample results would be if your initial conjecture is true (there is a .0054 probability of obtaining all 7's and 11's on the first five rolls).
4. Make a decision about whether you think that initial conjecture is true.

This procedure is exactly what you do with tests of significance. In fact, you will always decide how *significant* the evidence is against the conjecture, based on how likely the data are to occur when the conjecture is true. If the data are very unlikely to occur when the conjecture is true, this gives strong evidence against the conjecture/hypothesis.

What's a little bit frustrating is that you will never know for sure whether you are making the correct decision, just as you will never know for certain whether the confidence interval captures the parameter value. Still, you will be able to make some quantitative statement about how convincing the evidence is.

The second cumbersome component of trying to understand tests of significance lies in the volume of new terminology involved. However, it is important to remember that the choice of terminology and notation is very crucial in assessing your understanding of the overall process. For example, you aren't calculating the probability that the dice are fair. You started off assuming the dice were fair and then made a decision to keep this belief or reject it based on the probability that fair dice show such outcomes. This is very different from deciding how likely it is that the dice are fair.

In the following example, you will step through the process in detail. We have highlighted some places where students often veer down the wrong path.

Example 2: It has been conjectured that when asked whether they prefer to hear good or bad news first, people tend to choose bad news. Suppose we have a simple random sample of 95 Cal Poly students and find that 83% wanted to hear bad news first. Is this convincing evidence that a majority of all Cal Poly students prefer to hear bad news first?

Solution:

The data and parameter

In this study, the observational units are the students, and the binary variable is which type of news (good or bad) the student prefers to hear first. The sample is the $n = 95$ students, and the population is all students at Cal Poly. You want to know whether θ, the proportion of all Cal Poly students who prefer bad news first, is larger than .5. What you have observed is $\hat{p} = .83$, the proportion of the sample who prefer bad news first.

It is important to translate everything from the problem statement into the appropriate symbols. Defining the variable helps you determine whether the problem deals with means or proportions. It is also important to be clear about what is known (the value of $\hat{p}$) and what is unknown (θ, though you can define it in words). Keep in mind that you define the parameter (θ) to be a number, not the variable ("which type of news do students prefer to hear first") and not the research question ("Do Cal Poly students tend to prefer to hear bad news first?") and not the population (Cal Poly students).

Numerical and graphical summaries

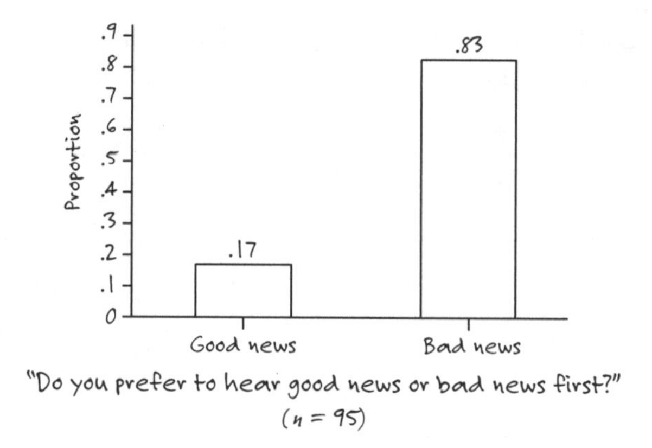

In this sample, a large majority indicated that they would prefer to hear bad news first.

Inference

The sample definitely had a tendency to want to hear bad news first, but was this a strong enough tendency to convince you that the population has the same tendency? You will start off assuming no and see whether the data convince you otherwise.

The hypotheses. The uninteresting case here is that there is no preference between the two choices. This would correspond to θ being equal to .5. However, before you even saw the data, it was suspected that people were more likely to prefer hearing bad news first, so you will conjecture that $\theta > .5$.

Translating these into hypothesis statements gives

H_0: $\theta = .5$ (there is no preference in the population for hearing bad news first)
H_a: $\theta > .5$ (a majority of Cal Poly students prefer hearing bad news first).

Note that this hypothesized value, .5 in this example, is often denoted by the symbol θ_0.

So, as in the earlier dice example, you will now choose between two options:

either

- $\theta = .5$ and $\hat{p}$ is larger than .5 in this sample by random chance

or

- $\theta > .5$.

You will decide between these two options by determining how likely it is to observe $\hat{p} \geq .83$ when θ is actually equal to .5. With the dice example, you had some intuition that rolling all 7's and 11's was a pretty unlikely result for fair dice. How do you know whether .83 is an unlikely value for $\hat{p}$ when $\theta = .5$? Assume the null hypothesis is true and $\theta = .5$ (just as you assumed the dice were fair). If $\theta = .5$, what could you say about typical values for $\hat{p}$ if you took different samples from this population? This is where the Central Limit Theorem comes into play. The Central Limit Theorem, if it applies, tells you about the pattern of variation in $\hat{p}$ values for different random samples.

Checking technical conditions (assuming H_0: $\theta = .5$ is true).

- You were told the sample was a simple random sample from all Cal Poly students, so you can consider the results to be representative of all Cal Poly students.
- $n\theta_0 = 95(.5) = 47.5$, and $n(1 - \theta_0) = 95(.5) = 47.5$. [*Remember:* You assume the null hypothesis $\theta = .5$ is true here and so you use the value of .5 in this check. With confidence intervals, you don't assume any value for the parameters and so you used the sample proportion in the check. With confidence intervals you make no initial statement about θ.] Because both values exceed 10, you can assume the sampling distribution is reasonably modeled by the normal distribution.

Both of the conditions are met, so the Central Limit Theorem applies. The Central Limit Theorem tells you that if you were to take many different random samples (of 95 students each) from a population with $\theta = .5$, then the resulting sample proportions would be centered at .5, the standard deviation would equal $\sqrt{(.5)(.5)/95} = .0513$, and the shape would be modeled by a normal distribution:

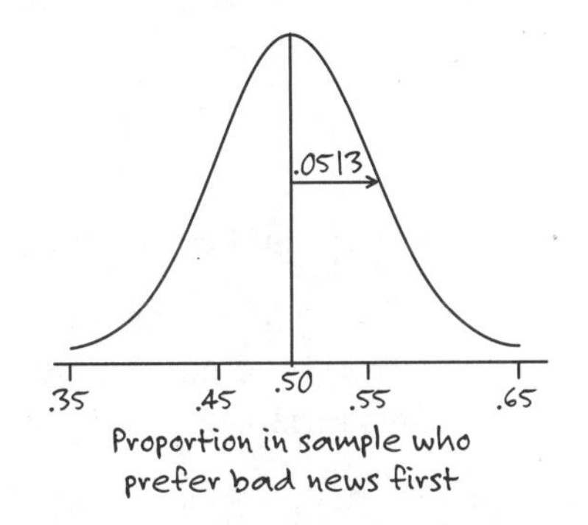

Test statistic and p-value. Now that you can describe typical values for $\hat{p}$, we can make a judgment about the value of $\hat{p}$ that you actually observed, .83. To see how far this value is from .5, standardize the observation:

$$z = \frac{.83 - .5}{.0513} = 6.43$$

This z-score indicates that .83 is more than six standard deviations from .5. That's extremely surprising!

The next step is to determine how often you get a z-value at least this extreme. Since you used the normal distribution as a model, refer to *WS* Table II. However, you find that $z = 6.43$ is way off the chart. So the best you can determine from the table is that the p-value is less than .0002 (the smallest probability listed in the table). You can express this symbolically:

$$\text{p-value} = \Pr(\hat{p} \geq .83 \text{ when } \theta = .5) \approx \Pr(Z \geq 6.43) < .0002$$

Conclusion

The calculations have indicated that it would be highly surprising to get a sample proportion greater than .83 if the population proportion had indeed been equal to .5. In fact, this is even more unlikely than only rolling 7's and 11's in five rolls. So now comes the decision:

> Do you think θ is .5 and the sample results were simply very unlikely, or do you think $\theta > .5$?

Often you can make a subjective judgment of whether this is convincing evidence to you. But if you want a cut-off value, common values are .10, .05, and .01. If the p-value is less than the cut-off, this means the data are surprising when H_0 is true and so you reject the null hypothesis (H_0). In this case, .0002 is far less than all common cut-offs, so you will reject H_0 and conclude that more than half of all Cal Poly students prefer to hear bad news first (i.e., that $\theta > .5$). You don't know for sure that $\theta > .5$, but you know it would be pretty shocking to get a sample proportion ($\hat{p}$) value as large as .83 if θ was .5. This provides very strong evidence that $\theta > .5$, indicating that, yes, a majority of Cal Poly students do prefer to hear bad news first.

KEY CONCEPTS

- The following list summarizes the steps and calculations for a test of significance:

  1. *Define the parameter of interest.* You should specify the correct symbol and what this symbol represents in words. It is often very helpful to start by considering what the observational units and variable are.

  2. *State the two hypotheses.* The null hypothesis is typically the "uninteresting" case, and in this text will always involve an equals sign. The alternative hypothesis is what the researchers are hoping to show, and in specifiying H_a you will need to choose between $<$, $>$, and $\neq$. It's often easier to start with the alternative, since it comes from the research question typically stated in the problem.

     Check the technical conditions and consider the behavior of the sampling distribution of the statistic. This step enables you to determine whether it is valid to perform the test procedure. It is important that you state *and* check the technical conditions before you start plugging numbers into formulas. (Merely placing a check mark beside the stated conditions is not good enough.) In the case of a test of significance for a proportion, you assume the null hypothesis is true for the entire calculation, so you use the hypothesized value of the population proportion in checking the technical conditions.

  3. *Calculate the test statistic.* By standardizing the observed statistic value, you have a measure of the distance between the observed value and the conjectured value. In calculating the standard deviation of $\hat{p}$, you use the hypothesized value of θ (recall that when you calculate confidence intervals, you don't typically have a hypothesized value for θ and so you use $\hat{p}$ to approximate the standard deviation).

4. *Calculate the p-value.* The p-value indicates how often you would observe a statistic at least this extreme. Use the alternative hypothesis to determine what is meant by "extreme" (which direction or both directions). Keep in mind that this calculation is done assuming that the null hypothesis is true, and that the size of the p-value indicates how strong the evidence is against the null hypothesis.
5. *Make a decision about the null hypothesis.* A small p-value constitutes evidence against the null hypothesis. If the p-value is less than the level of significance α, we reject H_0. If the p-value is not small, do not reject H_0. Keep in mind that this decision is always about H_0 (you don't "accept" or "reject" H_a). Statisticians actually consider the statements "accept H_0" and "fail to reject H_0" as meaning different things! So you must be careful to state only that you did not have enough evidence against the null hypothesis to get rid of it. This process is similar to what happens in a criminal trial: We start off believing the defendant is innocent. If the evidence against the defendant is strong, we call him/her guilty. If the evidence against the defendant is not strong, we don't say we know he is innocent, we just say "not guilty"—that there is not enough evidence to convict based on the data at hand. Similarly, in statistics we say we did not have enough evidence against the null hypothesis ("fail to reject H_0"), instead of saying the null hypothesis is true ("accept H_0"). Unlike a trial, though, in statistics, you can often gather more data and try again!
6. *State a conclusion about the research hypothesis in context.* Don't forget to go back and answer the original question.

21

- The goal of a test of significance is to evaluate the amount of evidence the sample provides against some claim about the population parameter.
- The general form of the null hypothesis is

  H_0: parameter = hypothesized value

  and the general form of the alternative is

  H_a: parameter $<$, $>$, or $\neq$ hypothesized value.

  Note that the null and alternative hypotheses can never be true at the same time (so don't include an equality option in H_a).
- The general form of a test statistic is

  $$\frac{\text{estimate} - \text{hypothesized value}}{\text{standard deviation of estimate}}$$

  - Larger values (in absolute value) of the test statistic provide more evidence against the null hypothesis. (In *WS* Activity 21-1, e.g., Celia's performance—her ability to convince others that she could distinguish between the colas—was more convincing than Brenda's.)

- Since the standard deviation, $\sqrt{\theta_0(1-\theta_0)/n}$, decreases if you increase the sample size, the test statistic,

$$z = \frac{p - \theta_0}{\sqrt{\theta_0(1-\theta_0)/n}}$$

becomes larger. A larger z-value produces a smaller p-value and therefore stronger evidence against the null hypothesis for the sample proportion $\hat{p}$, as shown in the following figures.

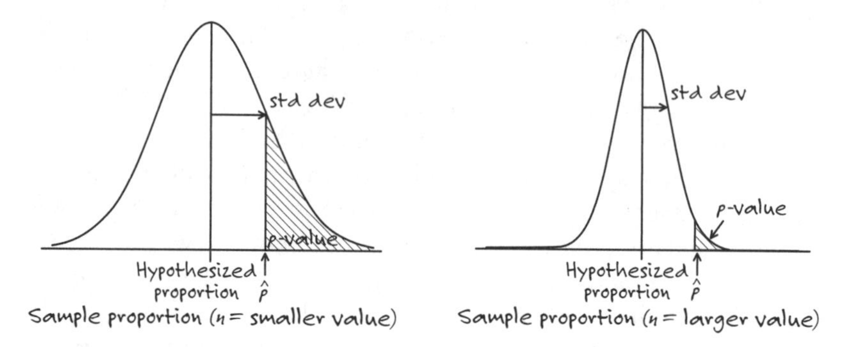

- The p-value indicates how often you would get data "like this" when the null hypothesis is true—that is, how often you would get a sample result at least this far from the conjectured value.
  - Small p-values provide strong evidence against the null hypothesis.
  - A large p-value indicates that it is fairly common to observe a sample value this close to the conjectured value when the null hypothesis is true, so that you do not have compelling evidence against the null hypothesis.
  - The p-value is always a "tail probability," that is, you want to know how often you would get a result at least this extreme (at least this far out in the tail).
  - You can't determine whether a sample proportion is "significantly" different from a conjectured value for the population proportion without knowing the sample size.
  - Given the same values of $\hat{p}$ and θ_0, larger samples provide stronger evidence against the null hypothesis than smaller samples. (However, the population size still doesn't matter.)

CALCULATION HINTS

- Try to break down the test statistic calculation into pieces. Carry a few extra digits and then round at the end (e.g., carry to the hundredths place before looking the z-value up in Table II).

- Include a sketch of the sampling distribution with the area representing the p-value shaded. Remember to shade in the direction specified by the alternative hypothesis. This will help you judge whether the p-value you calculate is a reasonable value.

- The basic steps that are common to all one-sample z tests for a population proportion are outlined below. The figure shows a sampling distribution of sample proportions.

H_0: θ = hypothesized proportion (θ_0)

Checking technical conditions

- Sample data are a simple random sample from the population of interest.
- $n\theta_0 \geq 10$ and $n(1 - \theta_0) \geq 10$.

Test statistic

$$z = \frac{\hat{p} - \theta_0}{\sqrt{\theta_0(1 - \theta_0)/n}}$$

p-value

You can find the p-value from Table II.

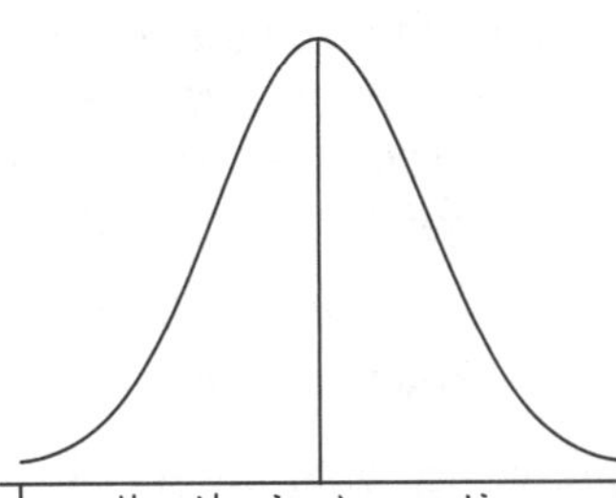

COMMON OVERSIGHTS

- Not describing the parameter as a number (e.g., describing the variable or the population itself instead).
- Not realizing (and not clearly specifying) that the hypotheses are statements about population parameters (as opposed to sample statistics).
- Stating the p-value as a probability statement about the parameter or the hypotheses.
    - The parameter is not "random;" for example, it doesn't make sense to use the language "the probability that θ equals .5 is . . . ." Similarly, the null hypothesis either is true or is not. You aren't making a probability statement about how likely it is that H_0 is true.
    - Also keep in mind that you should state the hypotheses *before* you see the sample data, solely based on the question specified by the researchers. Here are some guidelines:
      "Is there evidence of a decrease?" $\rightarrow H_a: \theta <$
      "Is there evidence that the proportion is now larger?" $\rightarrow H_a: \theta >$
      "Is there evidence of a difference?" $\rightarrow H_a: \theta \neq$
- Not presenting a thorough discussion of the technical conditions. Make sure you can differentiate between what the condition is and how you check it. Don't just say "check" without giving details (e.g., how do you know the condition is met; show the details on your check). It may not even be appropriate to continue with the calculations if the technical conditions are not met.
    - The p-value is only valid if the data are from a simple random sample and if n is large. If either of these conditions is not satisfied, you can't safely interpret the p-value.
- Not referring to the p-value when making your decision about the null hypothesis.
    - Make sure it is clear what evidence you are basing your decision on.

- Stating your conclusion in terms of "accepting the null hypothesis" or saying that the data "provide strong evidence in favor of the null hypothesis."
  - These statements are not appropriate because a test of significance only assesses the strength of evidence against the null hypothesis. Even a large p-value does not provide evidence that the null hypothesis is true. The calculations assumed the null hypothesis was true. It would be circular reasoning to then turn around and say this "proves" the null hypothesis.
- Not giving a final summary sentence in the context of the problem.
- Confusing the test statistic (z) with the p-value.
  - Remember the test statistic is the "ruler" for how far the sample result is from the hypothesized value. The p-value indicates how likely this is to happen by chance alone. You need to convert the test statistic into the p-value using Table II.
- Confusing $\hat{p}$ with θ or with the p-value.
  - You must be very careful with your notation, especially when stating the hypotheses.
- Confusing confidence level with significance level.
  - You will see in *WS* Topic 23 that these two ideas are closely related, but it is best to use these terms separately.
- Always using .5 as the hypothesized value instead of seeing what value the research question asks about.

WHAT WENT WRONG?

The following are examples of errors in the analysis of the example problem. In each case, explain what is wrong.

1. Checking technical conditions for the example with $95(.83) = 87.85$ and $n(1 - \theta) = 95(.17) = 16.15$.

2. Stating the hypotheses as $H_0 = .5$ vs. $H_a > .5$.

3. Stating the hypotheses as H_0: $\mu = .5$ vs. H_a: $\mu > .5$.

4. Stating the hypotheses as H_0: $\hat{p} = .83$; H_a: $\hat{p} > .83$.

5. The parameter is whether a student prefers to hear good or bad news first.

6. The p-value reveals that the probability that H_0 is true is essentially zero.

7. Expressing the conclusion as: "Reject H_0."

Solutions:

1. This presentation is technically incorrect because you want to do the check assuming the null hypothesis is true, which means you use the hypothesized value for θ, θ_0. This student used the sample proportion, $\hat{p}$.

2. This statement is wrong because the parameter does not appear anywhere in the hypotheses. The whole point of the test is to assess the plausibility of a hypothesized value for the parameter. In this statement, how do we know what is equal to .5? Remember, "H_0" is simply shorthand for "null hypothesis" and is not a parameter.
3. Statisticians will always assume you are indicating a mean if you use μ, but the parameter in this example is a population proportion, not a population mean.
4. The hypotheses always make statements about the population value, not the sample statistic. The $\hat{p}$ symbol is for the sample statistic. You know $\hat{p} = .83$; you don't need to test that! What you don't know is the value of θ, so the hypotheses need to state some possible values for θ.
5. This statement is a good description of the variable in this study, but it does not describe the parameter. To define the parameter, you have to describe a number. In this case a correct description of the parameter is the proportion of all Cal Poly students who prefer to hear bad news first.
6. This conclusion is very tempting, because a very small p-value such as this does indicate very strong evidence against the null hypothesis. The p-value is a probability, but about something random. What changes from sample to sample is the sample proportion. The null hypothesis is not sometimes true and sometimes not true. The p-value indicates the probability of obtaining such extreme sample data if the null hypothesis were true, which is *not* the same as the probability that the null hypothesis is true. This limitation is similar to the one that prohibits you from making a probability statement about a confidence interval.
7. This statement is correct but incomplete. *Always* relate the conclusion back to the context: The sample data provide very strong evidence that more than half of all Cal Poly students prefer to hear bad news first.

FURTHER EXPLORATION

Open the "Test of Significance Calculator" Java applet at *www.rossmanchance.com/applets/TOSCalculations/TOSCalculations.html.*

a. Change the direction of the inequality in H_a to match the bad news first example by clicking the < button, changing the direction of that button to >. Specify 95 as the sample size, and then .83 as the sample proportion. The applet will then determine the count (sample number of successes .79) for you. Click the **CALCULATE** button, and confirm the test statistic and p-value that you determined in the example.

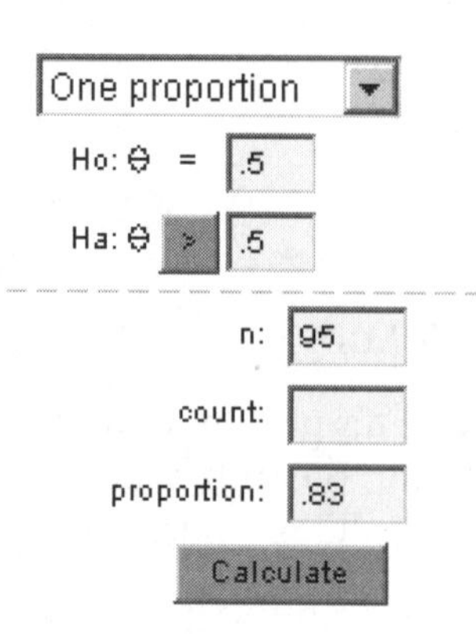

b. Suppose only 50 people in the sample preferred bad news first. Change the count to 50, and click the **CALCULATE** button. How does the test statistic value change? How does the p-value change? Why does this make sense?

c. Suppose 52.6% of the sample agreed, as in part b, but that the sample size had been 190 instead of 95. Enter this new value for n in the applet, and click the **CALCULATE** button. Again, comment on how the test statistic and p-value change. Why does this make sense?

d. Change the direction of the inequality in the alternative hypothesis to "<," and click **CALCULATE**. How is this p-value related to the p-value in part c?

e. Change the inequality in the alternative hypothesis to "<>" (which is the applet's symbol for $\neq$, not equal to), and click **CALCULATE**. How is the two-sided p-value related to the p-value in part c?

f. In a sample of 150 students, how many would have to prefer bad news first to convince you to reject the null hypothesis at the 5% level in favor of the original conjecture? Try to determine this by hand first. Then, in the applet, set the p-value to .05 (and the inequality in the alternative hypothesis back to ">"), and click the **REVERSE** button.

Solutions:

a. You should again obtain $z = 6.47$ and p-value ≈ 0.

b. The test statistic is much closer to 0 ($z = 0.51$) and the p-value is larger (p-value = .3061). The observation of 50 is not as many standard deviations from the expected value ($95 \times 5 = 47.5$). Since the observed count is much closer to the expected count when $\theta = .5$, the probability of observing a value at least this extreme when the null hypothesis is true is larger.

c. The test statistic is further from zero ($z = 0.72$) and the p-value is smaller (p-value = .2368), as the sampling variability will decrease with the larger sample size. Therefore, the same observed $\hat{p}$ value is less likely to happen by chance.

d. The new p-value is one minus the value of the old p-value, .7632.

e. The two-sided p-value, .4735, is twice the size of the one-sided p-value.

f. With a sample size of $n = 150$, the standard deviation of the sample proportion is $\sqrt{(.5)(.5)/150} = .0408$. For the p-value to be at most .05, the test statistic would need to be at least $z = 1.645$ (found by looking up a probability of .9500 in Table II, reading backward, and taking either candidate z-value or taking the midpoint between the two candidate values of 1.64 and 1.65). To be 1.645 standard deviations above the expected proportion, you calculate $.5 + 1.645(.0408) = .567$. Therefore, you will reject H_0: $\theta = .5$ in favor of H_0: $\theta > .5$ whenever $\hat{p} \geq .567$, which is when 84 or more in the sample of 150 prefer to hear bad news first.

Some things to note from this exploration:

- Changing the sample result to be closer to the hypothesized value (which is what you did in part b by making the sample proportion closer to .5) makes the test statistic closer to zero and, thus, the p-value larger. This indicates weaker evidence against the null hypothesis, which makes sense because the sample result is now closer to what the null hypothesis says.
- Increasing the sample size while leaving the sample proportion unchanged, as you did in part c, increases the evidence against the null hypothesis and so decreases the p-value. You would be less likely to get such an extreme result with a larger sample.
- If you change the direction of the inequality in the alternative hypothesis, the new p-value can be found by subtracting the old p-value from 1. (Remember that sketching the normal distribution is a good way to help you calculate the correct probability.) In part d, the sample result did not differ from the hypothesized value in the direction specified by the alternative, so the p-value turned out to be greater than .5. This is a very large p-value, which indicates no evidence against the null hypothesis and therefore no evidence in favor of this particular alternative hypothesis.
- Changing the inequality in the alternative hypothesis to "not equals" yields a two-sided the p-value, which is double the one-sided p-value, suggesting weaker evidence against the null hypothesis. This relation makes intuitive sense—if you don't suspect a particular direction, you will need slightly more evidence to convince you of a particular direction.

Topic 22:

TESTS OF SIGNIFICANCE II: MEANS

In this topic, you reviewed the general ideas behind a test of significance and applied it to quantitative data. Like the calculations for a confidence interval for a population mean, these calculations involve the *t*-distribution. Again, one of the first questions to ask is whether the data are categorical or quantitative. Then ask yourself whether you are assessing evidence against some claim about the parameter or estimating the parameter. If you are assessing evidence, then conduct a test of significance. If you are estimating a value, then construct a confidence interval.

Example: Recall the scenario in *WS* Activity 22-17. Researchers at Stanford University studied whether reducing children's television viewing might help prevent obesity. At the beginning of the study, children in the third and fourth grades were asked to report how many hours of television they watch in a typical week. The 198 responses had a mean of 15.41 hours and a standard deviation of 14.16 hours. Do these data provide evidence at the .05 level for concluding that third- and fourth-graders watch more than 2 hours of television per day on average?

Solution:

The data and parameter

The observational units are the students. The sample is the 198 students. How to define the population is a little less clear. For now, let's define it to be third- and fourth-graders. The variable measured, hours of television watched per week, is quantitative. So the parameter is μ = mean number of hours of television watched per week by third- and fourth-graders. Since you are working with the number of hours per week, watching more than 2 hours per day corresponds to watching more than 14 hours per week.

Numerical and graphical summaries

Unfortunately, you don't have access to the individual observations here, only the sample mean and sample standard deviation. The sample mean was greater than 14 hours, supporting the researchers' conjecture.

Inference

The hypotheses.

H_0: $\mu = 14$ (Children in this population watch an average of 2 hours of television per day, an average of 14 hours per week.)

H_a: $\mu > 14$ (Children in this population watch more than 14 hours of television per week, on average.)

You know that the alternative should correspond to *greater than* 14 because the problem asked whether there was evidence that children watch *more than* 2 hours per day on average.

Note that the hypothesized value for the population mean (14 in this example) is often denoted by the symbol μ_0.

Checking technical conditions.

- You should have serious doubts about whether this is a random sample from the population specified, but you can still ask how often you would get a sample mean this much greater than 14 by chance alone.
- You don't know anything about the population distribution of viewing times, but since the sample size is quite large ($n = 198 \geq 30$) you can consider the sample size/normal population condition to be met.

Therefore, you should apply the one-sample t-procedure.

Test statistic and p-value.

$$t = \frac{15.41 - 14}{14.16/\sqrt{198}} = 1.401$$

Using *WS* Table III with d.f. = 100 (rounding down from the exact value of 198 − 1 = 197) and looking for $t = 1.401$, you find that the p-value is between .05 and .10. Technology, such as the applet shown here, reveals the p-value more exactly to be .0814.

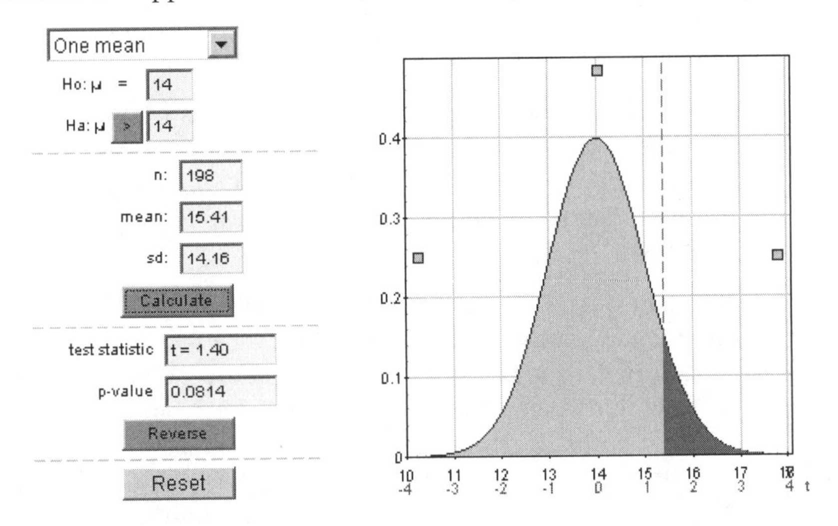

Conclusion

You were told to use the 5% significance level. Because the p-value .0814 is greater than .05, you fail to reject H_0. A sample mean at least this large would occur in about 8% of samples when $\mu = 14$, which is not a "statistically significant" difference at the 5% level. So, you do not have strong evidence that third- and fourth-grade students (from this population) watch more than 2 hours of television per day on average. You must be cautious in generalizing these results to a large population since the sample was not necessarily randomly selected.

KEY CONCEPTS

- Don't forget to check the technical conditions. Your check might involve graphing the sample data (especially when the sample size is small) to see whether it is reasonable to believe the population distribution follows a normal shape.
- You can't determine whether a sample mean is "significantly" different from a conjectured value for the population mean without knowing the sample size.
- With quantitative data, the standard deviation also plays a role in the p-value. If the distribution has more variability (larger standard deviation), the test statistic will be smaller and thus the p-value larger. So with more variability, it will be more difficult to detect a difference from the conjectured value.
- Although the text discussed confidence intervals first, statisticians often conduct the test of significance first. Then, if they reject the null hypothesis (e.g., decide that μ is not equal to 14), they follow up with a confidence interval to determine the plausible range of values (what is μ?).

- If you have taken two measurements on each person, as with a pre-test and post-test analysis, then the appropriate analysis is to calculate the difference in measurements for each person and then apply this one-sample t-procedure to those differences (as in *WS* Activity 22-4).

CALCULATION HINTS

- When calculating the p-value, to sketch the sampling distribution, you can jump straight to the distribution of the test statistic, which follows the t-distribution. Indicate where the test statistic falls and shade in the direction(s) specified by the alternative hypothesis.
- With the z-table you were able to determine probabilities and p-values pretty exactly. This is not true with the t-table. When using the t-table, you first look for where the t-test statistic falls in the appropriate degrees of freedom row, see which two values it falls between, and then make a statement such as "the p-value is between .05 and .10" or "the p-value is less than .001," based on the corresponding values at the top of the column (remembering to multiply by 2 for a two-sided test). Fortunately, this kind of range gives enough information to draw meaningful conclusions.
- If the degrees of freedom are not in the table exactly, we suggest rounding down. This will give you a slightly larger than actual p-value, making your decisions a little more conservative.
- Don't confuse "critical values" with "test statistics." With a confidence interval, you are given a confidence level (essentially a probability) and work backwards to determine how many standard errors (as determined by the z^*- or t^*-values) you want to use. This is the "critical value." With a test of significance, you calculate the test statistic from the data and then convert that to a probability (p-value). The text uses the notation z^* and t^* for critical values and z and t for test statistics.
- The basic steps that are common to all one-sample t tests of significance for a population mean are outlined below.

H_0: μ = hypothesized mean (μ_0)

Checking technical conditions

- Sample data are a simple random sample from the population of interest.
- Either the population values follow a normal distribution or the sample size is large ($n \geq 30$). If the technical conditions are met, then, assuming the null hypothesis is true, the sampling distribution of the test statistic is modeled by the t-distribution, as shown in the graph.

Test statistic

$$t = \frac{\bar{x} - \mu_0}{s/\sqrt{n}}$$

p-value

You can find the p-value from Table III with degrees of freedom $n - 1$.

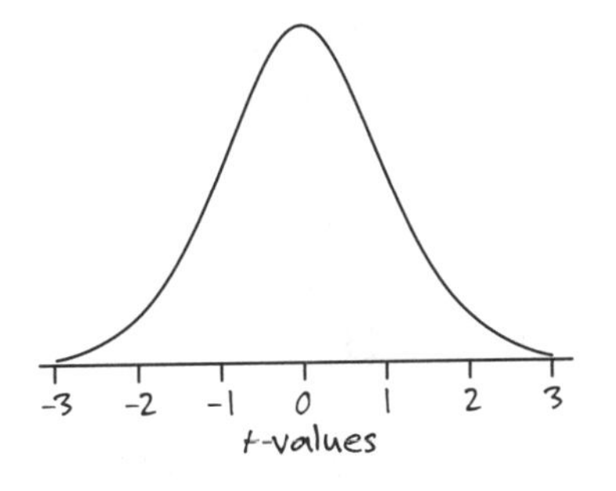

COMMON OVERSIGHTS

- Forgetting to first decide whether the data are quantitative or categorical and trying to apply the wrong formula. Finding that you are trying to plug a round peg into a square hole is a good indicator that you've made a mistake. For example, if you don't have a standard deviation to plug into a formula that needs one, you might be using the wrong formula.
- Forgetting to state the null and alternative hypotheses both in symbols and in words. Using both symbols and words helps ensure that you have interpreted the situation accurately. Also, don't forget the one-sentence summary at the end that places a statement about the population mean in context.
- Not realizing the t-test concerns a mean value and not individual observations. For example, you are not saying that no children watch more than 14 hours per week, just that there is not convincing evidence that the average amount they watch is not more than 14 hours.

WHAT WENT WRONG?

The following are examples of errors in the analysis of the example problem. In each case, explain what is wrong.

1. The parameter is the amount of time that children spend watching television per week.

2. The hypotheses are $H_0 = 14$ vs. $H_a > 14$.

3. The hypotheses are H_0: $\theta = 14$ vs. H_a: $\theta > 14$.

4. The hypotheses are H_0: $\bar{x} = 15.41$ vs. H_a: $\bar{x} < 15.41$.

5. The test statistic is $t = \dfrac{15.41 - 14}{14.16}$.

6. The test statistic is $t = \dfrac{15.41 - 2}{14.16/\sqrt{198}}$.

7. Because the p-value is greater than .05, we can conclude that the children in the population spend an average of 2 hours per day watching television.

8. If the p-value had been very small, we would have concluded that all children in the population watch more than 2 hours of television per day.

9. If the p-value had been very small, we would have concluded that most children in the population watch more than 2 hours of television per day.

Solutions:

1. This statement specifies the variable (the question "How much time do they spend watching television?") instead of the parameter (a number, the mean amount they watch per week).

2. These hypotheses do not make statements about the parameter. What equals or is greater than 14?

3. The θ symbol pertains to a proportion, but because the variable here is quantitative (amount of time spent watching television), the parameter is a mean, denoted by μ.

4. The hypotheses should always be statements about the unknown population parameter, μ, in this example. The symbol $\bar{x}$ is reserved for the *sample* mean and should not appear in a hypothesis statement. The observed value of the sample mean, 15.41 here, also should not appear in a hypothesis statement.

5. In this calculation the student forgot to divide by the square root of the sample size in the denominator, and so used the sample standard deviation instead of the standard error for $\bar{x}$.

6. In this calculation the student subtracted 2 in the numerator, perhaps thinking of hours per day instead of hours per week; the value to subtract is 14.

7. This is a tempting conclusion to draw, but it sounds too much like "accepting" the null hypothesis. Because the p-value was not terribly small here, you do not reject the null hypothesis, but that does not mean that you accept it. It is better to say that you do not have convincing evidence that children in the population spend more than an average of 2 hours per day watching television.

8. This conclusion is wrong because of the phrase "all children." A very small p-value would lead you to conclude that the *mean* time watching television exceeds 2 hours per day. But that the average exceeds 2 does not mean that every child's value exceeds 2.

9. The word "most" instead of "all" makes this conclusion more palatable than the previous one, but this statement is wrong for the same reason. Just because the mean value exceeds 2, you cannot conclude that most individual values exceed 2. Recall from your earlier study of the mean, that with a skewed distribution it's possible for a fairly small percentage of values to exceed the mean.

You will use the methods presented in Topics 21 and 22 if the goal is to assess the amount of evidence against a particular claim about a population parameter. You will recognize these problems by statements such as, "Is there evidence that...?" or, "Is the increase statistically significant?" You will then have to decide whether the variable of interest is quantitative (as in Topic 22) or binary (as in Topic 21). This determination will help you decide whether you are working with a proportion or an average, even if these words are not used explicitly. Remember to check the technical conditions before you start any calculations and that the conditions that need to be checked depend on whether you are working with quantitative or categorical data. You should clearly state which test statistic formula you are using and show the numbers plugged into the formula. When you are done with the calculation, use the p-value to state your conclusion about the null hypothesis and then include a one- or two-sentence summary that answers the research question in context (e.g., "There is evidence that . . .").

In calculating the p-value, you will shade beyond the sample result in the direction specified by the inequality in the alternative hypothesis. The following two graphs use the same sample result, but the p-value is calculated differently, depending on the direction of the inequality in H_a.

One-sided alternatives

H_a: parameter > hypothesized value

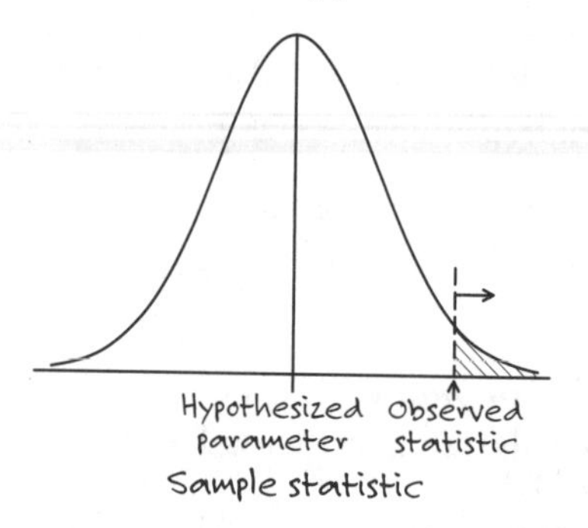

H_a: parameter < hypothesized value

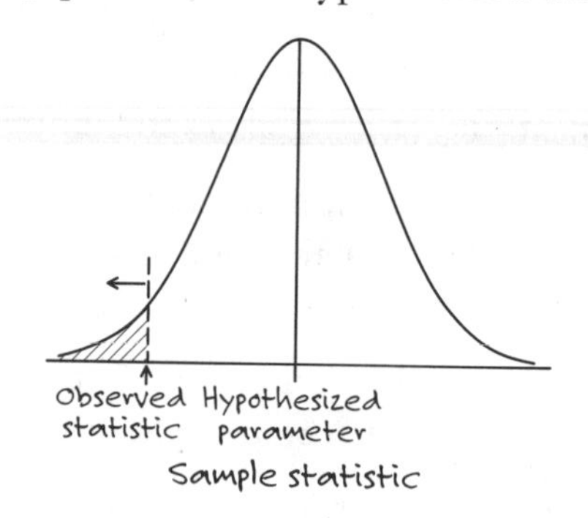

Two-sided alternative

H_a: parameter ≠ hypothesized value

When you don't have a prior conjecture about the direction in which the parameter value may differ from the hypothesized value, you must consider a result at least that extreme in either direction, so you state the alternative hypothesis with "≠" rather than ">" or "<." The p-value indicates how often you would have a sample result at least that far above the hypothesized population parameter *or* that far below the hypothesized population parameter if the null hypothesis is true. To calculate the p-value, find the probability in one tail and then multiply by 2.

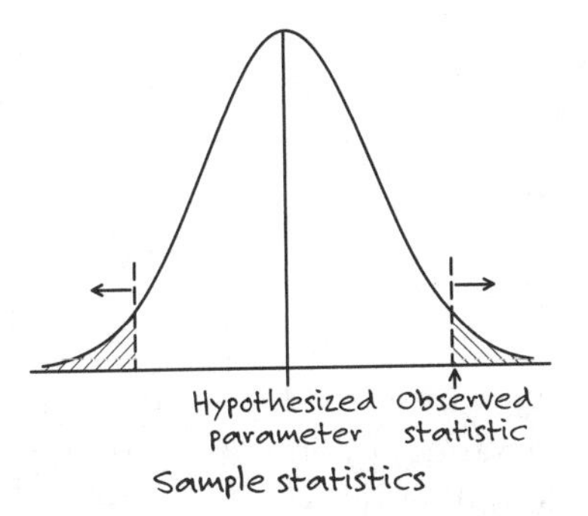

Note that if the sample result was in the opposite direction from what we hypothesized, then the one-sided p-value turns out to be greater than .5. Oh well!

H_a: parameter < hypothesized value, but sample result > hypothesized value

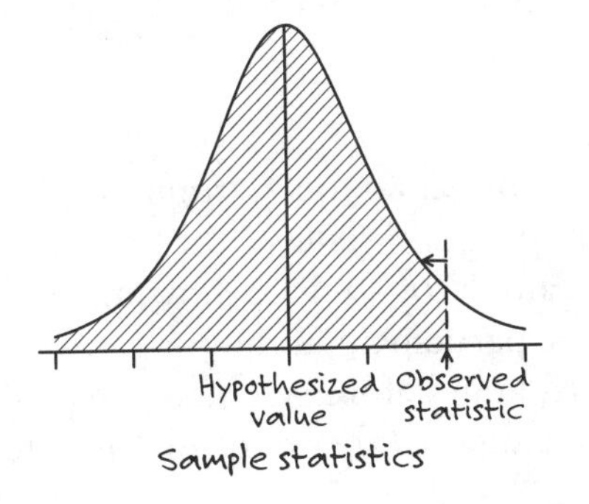

FURTHER EXPLORATION

Now that you have studied several inference techniques, you'll need to practice deciding which procedure applies to a given situation. For the following research questions, define the parameter and population of interest and indicate whether you would carry out a test of significance or a confidence interval calculation. If it is unclear from the information given, state what additional information you would need to make a decision. If you choose a test of significance, state the null and alternative hypotheses. If possible, check the technical conditions to see whether the procedure will be valid.

a. A pollster takes a random sample of 200 individuals and asks whether they think it is possible to live with a terrorist and not know it. The pollster wants to estimate the proportion of American citizens who think it is possible.

b. An English professor wants to decide whether a reading program decreases the amount of time it takes students to read a passage. He knows that in the past, students in his state averaged 10 minutes on the passage. He randomly selects 50 students who have participated in his program and gives them the reading exam.

c. A marketing manager wants to decide whether more than half of public schools (K–12) use Macintosh computers (rather than PCs). She examines a random sample of 100 public schools.

d. A campus administrator wants to determine whether participating in an alcohol abuse education program reduces the amount that college students drink. He asks a random sample of students how many drinks they have per month before and after they participate in the program. He calculates the differences in the amount of drinking before and after the program and wants to estimate the change in the average number of drinks per month.

Solution:

a. Let θ = proportion of American citizens who think it is possible to live with a terrorist and not know it. The goal here is to estimate θ, so you would calculate a 95% confidence interval for θ using $\hat{p} \pm z^* \sqrt{\hat{p}(1-\hat{p})/n}$. You are told the pollster takes a random sample. You don't know the sample proportion who think this is possible, so you can't check the other technical condition, but as long as $200\hat{p} \geq 10$ and $200(1-\hat{p}) \geq 10$ you could proceed.

b. Let μ = average amount of time to read the passage for students who have been through the reading program. The goal here is to decide whether $\mu < 10$, so you would perform a test of significance.

H_0: $\mu = 10$ (The average reading time after the program is the state average.)

H_a: $\mu < 10$ (The average reading time after the program is less than the state average.)

You are told the professor selects a random sample of students who have taken his program and since $n = 50$, the procedure should be valid whether or not the population distribution is normal. Still, you should look at a graph of the sample to make sure there are no severe outliers.

c. Let θ = proportion of all public schools who use Macintosh computers. You want to decide whether $\theta > .5$, so you would perform a test of significance.

H_0: $\theta = .5$ (Half the public schools use Macs.)

H_a: $\theta > .5$ (A majority of public schools use Macs.)

You are told the sample of public schools is random, and the other condition is met because $100(.5) = 50$ is greater than 10.

d. Let μ = average difference in number of drinks per month for all students at this college before and after the alcohol abuse education program. The goal is to estimate the value of μ, so you would find a confidence interval. You can use the formula $\bar{x} \pm t^* s/\sqrt{n}$ as long as the sample size is at least 30 (you should be very skeptical that the number of drinks per month is a symmetric distribution). You are told the sample was randomly selected, so that condition is met.

Topic 23:
MORE INFERENCE CONSIDERATIONS

This topic explored some of the finer points of inference procedures. Once you are comfortable with the mechanics and the basic reasoning process in the previous topics, then you should turn your attention to these important considerations of application and interpretation.

Example: Suppose you conduct an Internet poll and 3000 Internet users state whether they prefer Macintosh computers or PCs. You want to know whether the percentage of all computer users who use Macs is higher than the 5% you have heard reported.

a. If the population proportion of Mac users is actually larger than 5%, do you think the sample result will be statistically significant in this case? Explain.

23

b. Do you think it will be appropriate to apply the methods of this unit to answer this question?

c. Describe a type I and a type II error in this context.

d. How would a larger sample size change your answers to parts a–c?

Solution:

a. With a sample size as large as 3000, any sample proportion that is about 7% or higher will reject the null hypothesis that the population proportion equals 5% and lead you to conclude that more than 5% of this population uses a Mac. Whether this statistically significant difference is a practically significant question.

b. The sample size condition is certainly met, because 3000(.05) = 150 and 3000(.95) = 2850; both exceed 10, but you don't know how a simple random sample of all Internet users would be obtained. If you allow people to see the survey on the Internet, then those who reply to the survey are probably not representative of all Internet users and certainly not of all computer users. There is also a discrepancy between asking which brand they prefer and determining which brand they actually use.

c. A type I error would correspond to thinking the percentage of computer users who prefer Macs is higher than 5% in the population when actually it is not. A type II error would correspond to thinking that the percentage of computer users who prefer Macs is not higher than 5% when really it is.

d. A larger sample size will make it even easier for a result to be statistically significant but will not make the procedure any more valid if you are still allowing Internet users to voluntarily participate. The definitions of type I and type II errors will not change, but the probability of a type II error will decrease with the larger sample size.

KEY CONCEPTS

- If a two-sided test of significance rejects a conjectured value for the parameter, that parameter value should not be included in the corresponding confidence interval. By corresponding, we mean if a value is rejected at the α significance level, then it won't be in the $100(1 - \alpha)\%$ confidence interval, where α can vary. For example, if a value is rejected at the .05 significance level, then it will not appear within a 95% confidence interval. Similarly, any value not contained in a $100(1 - \alpha)\%$ confidence interval will be rejected as a hypothesized value by a two-sided test at the α significance level. Any value that is in the $100(1 - \alpha)\%$ confidence interval will not be rejected at the α level.
- The p-value tells you how strong the evidence is against the null hypothesis on a continuous scale, which is much more informative than the simple yes or no answer to the question about rejecting the null hypothesis.

- A test of significance procedure lets you know whether the parameter is a particular value, but an "it's different" answer doesn't tell you how different the parameter's value is. That is where the confidence interval comes in. Knowing the magnitude of the difference is especially important if the study has a very large sample size. With large samples, small differences are statistically significant even if you wouldn't consider them of importance in a practical sense.
- When you fail to reject the null hypothesis, there is always the chance you are making a type II error. This is one reason for our earlier caution against saying "accept H_0." Typically you control the probability of a type I error through setting α. If you have an α level in mind, you can often choose the sample size to try to achieve a certain minimal value for the probability of a type II error. This accommodation is very similar to determining the sample size necessary to achieve a certain width and confidence level in a confidence interval procedure.
- Always check a procedure's technical conditions and be very wary about generalizing conclusions from a sample not chosen randomly.

WHAT WENT WRONG?

The following are examples of errors in the analysis of the example problem. In each case, explain what is wrong.

Suppose that a student is told that when a coin was flipped 10,000 times, it produced 5,178 heads. The student decides to test whether this sample provides evidence that the probability of heads is not .5 with this coin. She calculates the test statistic to be $z = 3.56$ and the p-value to be about .0002.

1. If the student claims that the sample data provide strong evidence that the probability of heads is much larger than .5 for this coin, what's wrong with her conclusion?

2. If the student then calculates a 99% confidence interval for the probability to be (.497, .539), how might you know that she is mistaken without doing any calculations?

Solutions:

1. The error is in the phrase "much larger than .5." The small p-value does provide very strong evidence that the probability differs from .5, but with such a huge sample of 10,000 flips, that difference need not be large. In fact, a 95% confidence interval for the true probability of heads for this coin is (.508, .528). All values in this interval are larger than .5, but they are all fairly close to .5. Thus, you can reasonably conclude that the probability of heads is larger than .5 but *not much* larger than .5.

2. Since the two-sided p-value is less than .01, the 99% confidence interval should not include the value .5, which this incorrect interval does.

Unit VI:

Inference from Data: Comparisons and Relationships

Topic 24:
COMPARING TWO PROPORTIONS

In this topic you continued to practice the logic of statistical inference (tests and confidence intervals) while extending them to more involved data sets consisting of two or more groups or two variables. As you work through this topic, it is important to keep in mind the general ideas of significance and confidence that you learned in Unit V. In *WS* Topics 24 and 25, all that really changed was the tools used to calculate the test statistic. As you review these topics, try to focus on the settings that lead to each set of techniques: how many variables you have, what types of variables you have, and how many groups you are comparing.

Example: The example in *TK* Topic 7 included data on individuals who did and did not take a flu vaccine and compared whether they developed a flu-like illness. Of the 1000 people who got the flu vaccine, 149 of them still developed a flu-like illness. Of the 402 people who did not get the vaccine, 68 got a flu-like illness. In the earlier analysis you saw that although the difference was small, it was in the conjectured direction (fewer flu-like illnesses among those who took the vaccine).

a. Produce numerical and graphical summaries of these data. Discuss what they reveal.

24

b. Are the technical conditions met for you to carry out a two-sample z-test on these data?

c. Carry out the two-sample z-test: State the hypotheses, calculate the test statistic and p-value, and state your conclusion in context.

d. Calculate a 90% confidence interval for the difference in the population proportions between the two groups.

Solution:

a. The following segmented bar graph compares the vaccinated group to the nonvaccinated group:

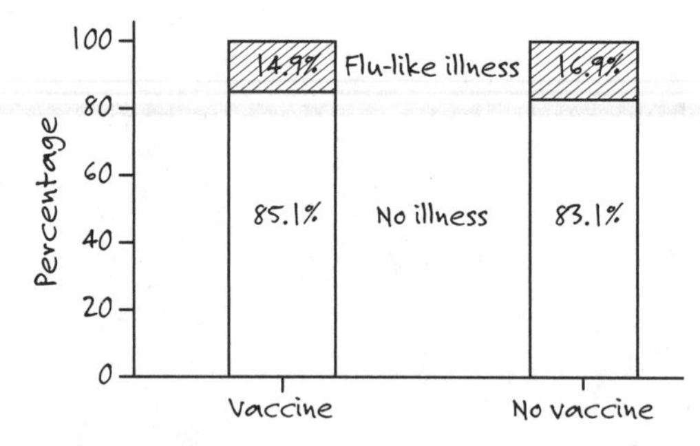

You can see that the vaccinated group was slightly less likely (14.9% versus 16.9%) to develop a flu-like illness.

b. No information was supplied as to how these individuals were chosen. It is clear the hospital workers decided for themselves whether or not they would take the vaccine, so this is not an experiment. It is not clear whether the workers studied are representative of all workers at the hospital, and you would certainly have doubts about generalizing these results to non-hospital workers. Thus, the random sample condition is not met.

The sample size condition is easily met given that the number of successes and failures in each group (149, 851, 68, 334) all exceed 5.

c. If you did carry out the test:

Let θ_1 = the proportion of all hospital workers who would develop a flu-like illness if they were to take the vaccine. Let θ_2 = the proportion of all hospital workers who would develop a flu-like illness if they were not to take the vaccine.

H_0: $\theta_1 = \theta_2$ (There is no difference in the population proportions who develop the illness.)
H_a: $\theta_1 < \theta_2$ (A smaller proportion of vaccine takers develop the illness.)

Observed: $\hat{p}_1 = \dfrac{149}{1000} = .149 \quad \hat{p}_2 = \dfrac{68}{402} = .169$

Combined sample proportion $\hat{p}_c = \dfrac{149 + 68}{1000 + 402} = .155$

$$z = \frac{\hat{p}_1 - \hat{p}_2}{\sqrt{\hat{p}_c(1 - \hat{p}_c)\left(\dfrac{1}{n_1} + \dfrac{1}{n_2}\right)}}$$

$$= \frac{.149 - .169}{\sqrt{.155(1 - .155)\left(\dfrac{1}{1000} + \dfrac{1}{402}\right)}} = \frac{-.02}{.02137} = -.94$$

p-value = $\Pr(Z \leq -.94) = .1736$ (from *WS* Table II)

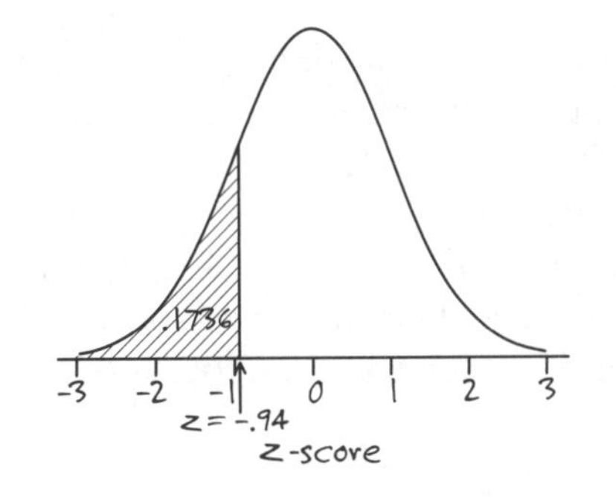

If there were no difference between these population proportions, you would find that the second sample proportion is at least this much larger than the first sample proportion in about 17% of samples. With this large p-value, you fail to reject the null hypothesis. You conclude that the proportion developing flu-like illness among those taking the vaccine is not significantly smaller than the proportion among those not taking the vaccine.

24

This conclusion is consistent with the discussion in *TK* Topic 7; the difference in the sample proportions does not seem substantial. Still, the sample sizes are rather large, so the test was necessary to determine whether the difference was statistically significant. You cannot, however, say that the flu vaccine did not work, only that you did not find compelling evidence to conclude that it reduced the proportion who developed a flu-like illness in the population.

You must be cautious, though, in believing that these samples are representative of all individuals who might take the vaccine. Perhaps hospital workers have a healthier lifestyle in general and are less likely to develop a flu-like illness and less likely to benefit from the vaccine than non-hospital workers.

Even if you had found a statistically significant difference, you would not be able to attribute that difference exclusively to the vaccine, since this was not a randomized experiment. Other confounding variables could be differentiating the two groups. Perhaps people who were concerned enough to receive the vaccine also modified their diet and took other health-related precautions.

d. For a 90% confidence interval, $z^* = 1.645$. Plugging that value into the formula in *WS* Activity 24-2, you find

$$(.149 - .169) \pm 1.645\sqrt{\frac{.149(1 - .149)}{1000} + \frac{.169(1 - .169)}{402}} =$$
$$-.02 \pm .036 = (-.056, .016)$$

You are 90% confident that the difference in the proportion who develop a flu-like illness is between .016 higher for the population receiving a vaccine to .056 higher for the population not receiving the vaccine. This interval also confirms that 0 is a plausible value for the difference in the population proportions at the 90% confidence level. Again, you must be cautious in carrying out this procedure since these were not random samples.

KEY CONCEPTS

- The reasoning of the test and the definition of the p-value when comparing the proportions of two groups is the same as in earlier topics: how often you would observe a sample result at least this extreme if the null hypothesis is true. When you look at the difference in the sample proportions, you assume the samples came from populations with the same probability of success. Also remember that if the difference in sample proportions is equal between two studies, the study with larger sample sizes will lead to a smaller p-value and therefore more evidence that the population proportions differ.
- The logic of the confidence interval with two proportions is also the same, including the interpretation of the confidence level and the effects of sample size and confidence level on the width of the interval. Just remember that when comparing the proportions of two groups the interval estimates the *difference* in population proportions between the two groups.

- It is important to recognize when a research question involves the comparison of two population proportions. One way to help you decide this is to ask yourself how many variables you have and whether they are quantitative or categorical.
- It is crucial to check whether the technical conditions are satisfied before you decide to carry out a procedure. It is not appropriate to carry out the inferential procedure if the conditions are not met. If you have doubts about the validity of the technical conditions, then you should include big red flags in your discussion of the conclusions that follow from the analysis.
- No matter how small the p-value is or how far the confidence interval is from 0, a difference between two population proportions does not automatically imply a cause-and-effect relationship. (A cause-and-effect conclusion can only be drawn from a randomized comparative experiment.)

CALCULATION HINTS

- As always, show the details of your calculations, and try to perform the calculations in small pieces. You should carry out to a few extra places along the way and round only at the end of the calculation.
  - Remember to work with proportions, not counts or percentages.
- Make sure your final conclusions are consistent with your initial numerical and graphical exploration of the data.
- When interpreting the confidence interval for $\theta_1 - \theta_2$, it is often most interesting to determine whether the interval contains 0 (because an interval that contains 0 indicates that there is no difference between the two population proportions).
- In general, it does not matter which group you label Group 1 and which you label Group 2, but it is important that you use the labels consistently throughout the problem. If someone else labeled them in the other order, then your test statistic would be the negative of theirs, and the endpoints of your confidence interval would be the negatives of theirs, but all of your interpretations would be exactly the same.

24

COMMON OVERSIGHTS

- Not choosing the correct procedure.
- Calculating the standard error term incorrectly; for example, switching the standard error formulas of the test statistic and the confidence interval.
  - Calculating the combined sample proportion $\hat{p}_c$ as the average of the two sample proportions. You can only use the average if the sample sizes are the same for the two groups.
- Not fully checking the technical conditions and not realizing the impact of violation of these conditions on the validity of interpreting the test results.
  - For example, considering only the sample size and not considering the proportion of successes when checking the sample size condition.
- Using symbols for samples statistics (e.g., $\hat{p}$) instead of population parameters in the null and alternative hypothesis statements.

- Specifying the null and alternative hypotheses based on what you observe in the data instead of the statement of the research question.
  - Incorrectly specifying one- versus two-sided alternative hypotheses and/or not being consistent about "sidedness" when you calculate the p-value.
- Not realizing that the confidence interval is for the *difference in the population proportions.*
- Not following up your calculations with an interpretation in context.

WHAT WENT WRONG?

The following are examples of errors in the analysis of the example problem. In each case, explain what is wrong.

The 90% confidence interval for $\theta_1 - \theta_2$ was determined to be $(-.056, .016)$.

1. This indicates that 90% of the sample proportions are contained in this interval.

2. This indicates that we are 90% confident that between 1.6% and 5.6% of the population developed a flu-like illness.

Solutions:

1. The confidence interval aims to specify $\theta_1 - \theta_2$, the unique difference in the population proportions. It is not a confidence interval for a difference in sample proportions or for a single proportion.
2. The confidence interval here is for the difference in the proportions of success, not for the overall proportion of success or for just one of the populations. That is why this interval may contain both positive and negative values. Negative values correspond to $\theta_2 > \theta_1$, and positive values correspond to $\theta_2 < \theta_1$.

FURTHER EXPLORATION

A survey of a college statistics class revealed 7 of the 19 women and only 4 of the 7 men preferred Pepsi® over Coke® when asked to choose between the two. Does the following output allow you to conclude that males and females are not equally likely to prefer Coke as they are Pepsi?

```
gender     X     N     Sample p
1          4     7     0.571429
2          7     19    0.368421

Estimate for p(1) - p(2): 0.203008
95% CI for p(1) - p(2): (-0.222950, 0.628965)
Test for p(1) - p(2) = 0 (vs not = 0): Z = 0.93 P-Value = 0.350
```

Solution:

This output can't be used! The technical conditions for the two-sample z-procedure are not met since you do not have at least five successes and five failures in both groups. Technology may provide results even when technical conditions are badly violated—it's up to you to check the validity of the procedure. Also, be careful with your wording. This analysis is not concerned with whether Coke and Pepsi are equally likely to be preferred, but whether the preference for women (proportion of women who choose Coke over Pepsi) is the same as the preference for men (proportion of men who choose Coke over Pepsi).

Topic 25:
COMPARING TWO MEANS

This topic paralleled *WS* Topic 24 and provided tools for comparing two population means. As in the one sample case, when working with quantitative data, you used the *t*-distribution to obtain critical values for confidence intervals and p-values. You should keep in mind which factors affect these calculations, such as sample size and the amount of variability within the samples themselves.

Example: Students in a statistics class took part in the memory experiment described in the Preliminaries of *WS* Topic 13 and in *WS* Activity 25-12. The following sample data are the number of consecutive letters correctly remembered before the first mistake for the two different arrangements of the letters.

Familiar three-letter packets (e.g., *JFK*):

6, 6, 6, 8, 9, 9, 9, 9, 12, 15, 15, 15, 15, 18, 18, 18, 19,
21, 21, 21, 21, 21, 21, 21, 24, 27, 27

Unfamiliar packets (e.g., *JFKC*):

2, 3, 3, 3, 5, 6, 6, 6, 6, 8, 9, 9, 10, 13, 14, 14, 14, 14, 14, 15, 15, 15, 17, 18, 20, 24

Carry out an analysis of these data to determine whether there is statistically significant evidence that the familiar three-letter packets improve recall.

Solution:

The first step in the analysis is to construct numerical and graphical summaries to compare these two sets of data. Since the variable is quantitative, you can use histograms, stemplots, dotplots, and/or boxplots to graph these data. Dotplots are shown here. Note that they have been drawn using the same horizontal axis scale, which is very important when trying to directly compare distributions.

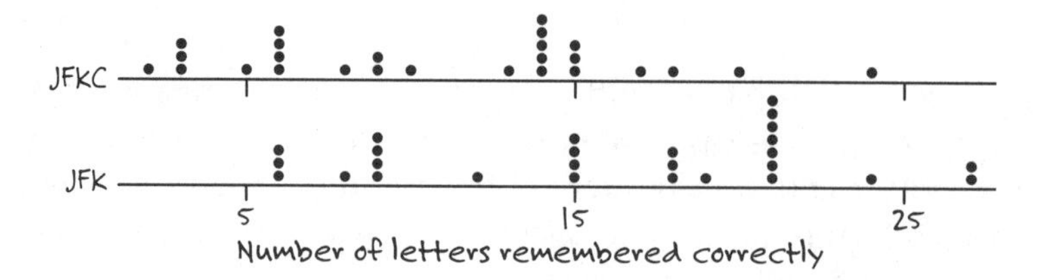

This graph provides informal evidence that the *JFK* distribution reveals a tendency for higher memory recall scores because its distribution is slightly shifted to the right of the *JFKC* distribution. The spread of the two distributions is similar. It is difficult to classify the shapes of these distributions because they are rather granular and irregular. The *JFK* distribution, especially, has many observations that are multiples of 3, which makes sense because those subjects were remembering the letters in groups of 3. There are not any extreme outliers, though the *JFKC* group did have one person who remembered 24 letters, which was high for that group.

If you compare the mean number of letters recalled, the *JFK* group shows a higher average: 16 letters versus 10.88 letters. Similarly, the median for the *JFK* group is higher: 18 letters versus 11.50 letters.

The standard deviations (6.45 letters for *JFK* versus 5.86 letters) and interquartile ranges (12 letters versus 9 letters) are similar, with slightly more variability in the *JFK* group.

There are no outliers according to the 1.5IQR criterion.

From this descriptive analysis, there is some evidence that those who receive the letters in familiar groups will remember about 6 more letters on average than those who receive the letters in unfamiliar groups. Because you are comparing two groups on a quantitative variable, consider using the two-sample *t*-test to see whether the difference observed in the sample means is statistically significant.

Let μ_1 = the mean number of letters recalled correctly for the population who could receive the letters in familiar three-letter packets.

Let μ_2 = the mean number of letters recalled correctly for the population who could receive the letters in unfamiliar packets.

H_0: $\mu_1 = \mu_2$ (There is no difference in the mean number of letters recalled.)

H_a: $\mu_1 > \mu_2$ (People who view the letters in familiar groups have better recall on average.)

Before applying the two-sample *t*-test, you must consider the technical conditions.

1. Are the two samples independently selected random samples from the populations of interest?

These are not really random samples, but this was a randomized experiment so you can consider the samples independent and you can use the two-sample *t*-test to see how often you would get a difference this big from the randomization process alone. However, you should not generalize your conclusions to a larger population.

2. Is the population normal or the sample size large?

   You have 27 and 26 observations in the two groups, so these do not quite satisfy the recommendation of at least 30 subjects in each group. But this is close. Looking at the dotplots, you see that these aren't the most symmetric distributions but there is no evidence of strong skewness or extreme outliers. So you can tentatively proceed since the sample sizes are reasonably large and the sample distributions are not too nonnormal.

Calculating the test statistic you obtain:

$$t = \frac{16 - 10.88}{\sqrt{\frac{6.45^2}{27} + \frac{5.8\,6^2}{26}}} = 3.03$$

You will use min(27 − 1, 26 − 1) = 25 as the degrees of freedom, and the t_{25} distribution is shown in the graph.

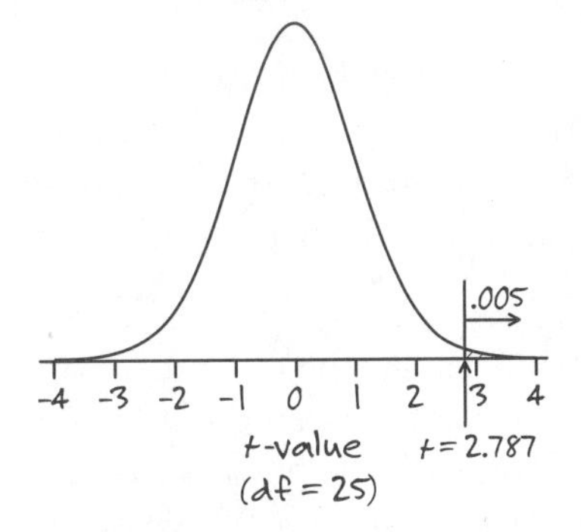

Using *WS* Table III you determine that the p-value is between .001 and .005. (Note that a t_{25} value of 2.787 has probability .005 lying to the right.) Technology tells us the p-value is .0028.

With such a small p-value (i.e., it is much less than .005), there is very strong evidence that the difference in the sample means did not happen by chance alone. You conclude that there is a real increase in the average number of letters recalled by those receiving the letters in familiar groups compared to the average number of letters recalled by those receiving the letters in unfamiliar order. You can draw this cause-and-effect conclusion because this was a randomized comparative experiment. However, you should be reluctant to generalize these results beyond the statistics students that participated in this experiment since they were not randomly selected from a larger population.

You can estimate the size of the difference by using a confidence interval for $\mu_1 - \mu_2$. A confidence level was not specified, so you can use 95%. The two-sample

t-interval has the same technical conditions as those checked earlier. The 95% critical value with d.f. = 25 is $t^* = 2.060$. So the interval becomes

$$(16 - 10.88) \pm 2.060\sqrt{\frac{6.45^2}{27} + \frac{5.86^2}{26}} = 5.12 \pm 3.48.$$

You can be 95% confident that those receiving the letters in familiar groups will remember between 1.64 and 8.60 more letters on average than those receiving the letters in unfamiliar groups.

KEY CONCEPTS

- The reasoning of the test and the definition of the p-value when comparing the means of two groups is the same as in earlier topics: how often you would observe a sample result at least this extreme if the null hypothesis is true. When you look at the difference in the sample means, you assume the samples came from populations with identical means. The size of the p-value will depend on the sample sizes involved, the magnitude of the difference in the sample means observed, and the sample standard deviations. If there is more variability within each sample, then the same difference in sample means will not seem as surprising as when there is less variability within each sample. The same difference in sample means will seem more surprising coming from large samples than from small samples.
- The logic of the confidence for the difference in two means is also the same, including the interpretation of the confidence level and the effects of sample size and confidence level on the width of the interval. Just remember that with the means of two groups the interval estimates the *difference* in population means between the two groups.
- As always, it is important to consider the technical conditions before you apply your inference procedure. You have been given the $n \geq 30$ guideline, but it is not a hard and fast rule. You need to consider the behavior of the data. The more nonnormal the sample looks, the larger the sample size you will require to confidently apply the procedure. If you have any questions about the validity of the procedure you should not carry out the inferential procedure or should at least discuss your hesitancy.
- It is also important to notice that these two-sample procedures (the subjects of Topics 24 and 25) are appropriate when you have a randomized experiment as well, even if you don't have a true random sample to begin with. Just realize that although a two-sample test will help you see whether the treatment effect is statistically significant, it will not enable you to generalize that conclusion to a larger population unless you did take random samples.
- You will often have two sets of quantitative data, but they are actually repeat measurements on the same set of observational units (e.g., a pre-test and a post-test). In this paired situation, a two-sample procedure is not appropriate. Instead, you should calculate the differences between these two measurements and apply a one-sample procedure to those differences, as in Topic 22.

CALCULATION HINTS

- In calculating the standard error of $\bar{x}_1 - \bar{x}_2$, remember that if you were told the standard deviations of each group, you need to square these quantities when substituting them into the formula.
- Recall that if the degrees of freedom you want does not appear exactly in Table III, we suggest you round down to the next d.f. listed in the table.
- Make sure your final conclusions are consistent with your initial numerical and graphical exploration of the data.
- When interpreting the confidence interval for $\mu_1 - \mu_2$, it is often interesting to determine whether the interval contains 0, which indicates that there is no difference between the two population means. You should also be prepared to test whether the difference in population means is something other than zero as well by subtracting that hypothesized value in the numerator of the test statistic.
- If you are carrying out a test involving a two-sided alternative, you will need to multiply the probability range that you obtain from Table III by 2. In the examples, you would have reported the two-sided p-value to be between .002 and .010.
- Once again it does not matter which group you label Group 1 and which you label Group 2 as long as you are consistent throughout the problem. If someone else defined them in the other order, then your test statistic would be the negative of theirs and your confidence interval endpoints would be the negatives of theirs, but all of your interpretations would be exactly the same.

COMMON OVERSIGHTS

- Not choosing the correct procedure.
- Not fully checking the technical conditions and not realizing the impact of violation of these conditions on the validity of interpreting the test results.
    - Not plotting the sample data to decide whether the distributions are reasonably symmetric and without extreme outliers.
- Using symbols for samples statistics (e.g., $\bar{x}$) instead of population parameters in the null and alternative hypothesis statements.
- Specifying the null and alternative hypotheses based on what you observe in the data instead of the statement of the research question.
    - Incorrectly specifying one- versus two-sided alternative hypotheses and/or not being consistent about "sidedness" when you calculate the p-value.
- Not realizing that the confidence interval is for the *difference in the population means.*
- Not following up your calculations with an interpretation in context.

WHAT WENT WRONG?

The following are examples of errors in the analysis of the example problem. In each case, explain what is wrong.

1. Since the interval does not contain 0, we are 95% confident that the scores for the first group are larger than the scores for the second group.

2. Those receiving the letters in familiar packets will perform better 95% of the time than those receiving the letters in unfamiliar packets.

3. The p-value of .003 indicates that there is a .003 probability that those receiving the letters in familiar packets will not perform better than those receiving the letters in unfamiliar packets.

Solutions:

1. We are 95% confident that the *difference in population means* is greater than 0. It is important that your interpretation does not sound like all values in the first population are larger than all values in the second population. Of course, it is also best to spell out what you mean by "first group" and "second group." In particular, it should be clear that the confidence interval estimates the difference in *population* means or the true treatment effect; it does not merely pertain to the samples in this study.
2. The "95%" of the confidence interval does tell you that something will be true "95% of the time." The "of the time" refers to repeating the random process many times and observing the resulting differences in sample means. Your conclusion here is either that the average recall score with the familiar packets is higher or it is not, not that sometimes it is higher and sometimes it is not higher.

3. This interpretation reads too much like the p-value is a probability statement about whether the null hypothesis is true. The p-value tells you the probability of observing a difference in sample means at least this large *assuming the null hypothesis is true*. There is a .3% chance that the difference in mean recall scores between these sample groups would be larger than 5.12 if there was actually no treatment effect.

FURTHER EXPLORATION

A group of students at a large college conducted a study on the weight of student backpacks and whether the student reported any back pain. Instead of only looking at backpack weight, they looked at the ratio of the backpack weight to the students' self-reported body weight (the body weight ratio). Based on the following output, is there evidence at the 5% level that the average body weight ratio for students who reported back pain is larger than the average body weight ratio for students who did not report back pain?

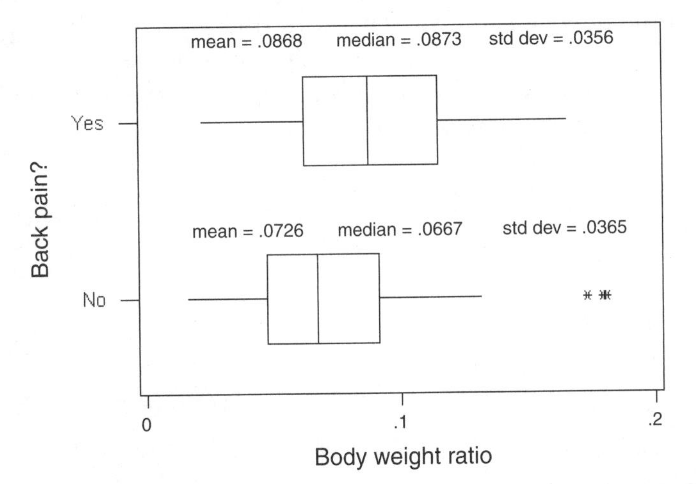


```
Two-sample T for Body weight ratio
back pain       N        Mean      StDev     SE Mean
No             68      0.0726     0.0365      0.0044
Yes            32      0.0868     0.0356      0.0063
Difference = μ (No ) - μ (Yes)
Estimate for difference: -0.01429
95% CI for difference: (-0.02967, 0.00109)
T-Test of difference = 0 (vs not =): T-Value = -1.86
P-Value = 0.068 DF = 62
```

25

Solution:

Before interpreting the computer output, you should consider the technical conditions. Unfortunately, you do not have any information on how these subjects were selected. If they were selected at random, then you could consider these to be independent random samples from the populations of those with back problems and those without back problems at this university. We cannot tell much about the shape of the distributions based on the boxplots, but with the large sample sizes we can safely apply the *t*-procedure as long as there is not extreme skewness or outliers. The sample without back pain does have a few outliers, which we should investigate to make sure the data were correctly recorded, although it does not appear they would be outliers in the group that has back problems. Since the outliers are not dramatically extreme, you can cautiously proceed.

The output shows that the *two-sided* p-value is .068. The corresponding one-sided p-value would be .032 (since the means are in the direction conjectured). This p-value is less than .05, so you would reject H_0: $\mu_1 = \mu_2$ (there is a difference in the mean body weight ratios of the population of students with back pain and the population of students without) at the 5% level in favor of H_a: $\mu_1 > \mu_2$ (the mean body weight ratio for the population of students with back pain is greater than the mean body weight ratio for the population of students without back pain).

Topic 26:
INFERENCE FOR TWO-WAY TABLES

In this topic you extended your descriptive analysis of two-way tables from *WS* Topic 7 to the chi-square test for assessing the significance of the relationship between two categorical variables. The chi-square procedure also allows you to test the equality of two or more population proportions. Even though the calculations in this topic appeared more complicated, you should keep in mind that the basic structure of a test of significance (stating the hypotheses, checking the technical conditions, determining a test statistic and p-value, and stating your conclusion in context) remains the same.

Example: A random sample of 93 statistics students at a large university were asked which lifetime achievement they would choose from the following three choices:

- To win an Olympic gold medal
- To win a Nobel Prize
- To win an Academy Award.

The instructor wanted to know if male and female students would respond differently. He noted that of the females, 15 chose the Academy Award, 29 the Nobel Prize, and 18 the Olympic gold medal. Of the males, 7 chose the Academy Award, 11 the Nobel Prize, and 14 the Olympic gold medal.

a. Is this an observational study or an experiment? Explain.

b. Identify the explanatory variable and the response variable. For each variable, specify whether it is quantitative or categorical.

26

c. Create a two-way table to summarize these data. Use the columns for the explanatory variable.

d. Create numerical and graphical summaries to display these data. Summarize what they reveal.

e. Would it be appropriate to carry out a chi-square test on these data? Explain.

f. Carry out the chi-square test, and state your conclusions. Use .05 as the level of significance.

Solution:

a. This is an observational study (a survey) since we did not impose the genders on the individuals.

b. The explanatory variable in this study is the individual's gender, which is a categorical variable. The response variable is which lifetime achievement the individual chose, also a categorical variable.

c. The two-way table with gender as the column variable looks like this:

	Females	Males	Total
Academy Award	15	7	22
Nobel Prize	29	11	40
Olympic gold medal	18	14	32
Total	62	32	94

d. The following segmented bar graph compares the two groups:

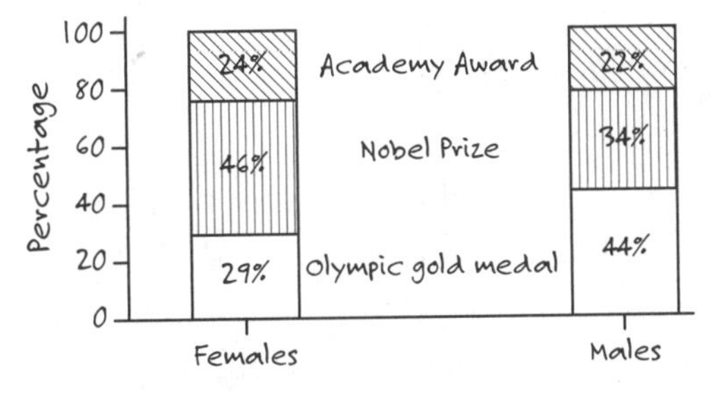

You can see that males were more likely to pick the Olympic gold medal than females (44% to 29%) whereas females were more likely to pick the Nobel Prize than males (46% to 34%). A similar percentage in the two groups (24% and 22%) picked the Academy Award.

e. In collecting these data, one sample was identified and two categorical variables were recorded about each person in the sample. The investigation concerns whether there is a relationship between gender and choice of lifetime achievement, which is what a chi-square test assesses. Next you must check the technical conditions for a chi-square test:

1. You were told the observations were a random sample selected from all statistics students at a large university.
2. When you compute the expected counts for each cell, you find:

	Females	Males	Total
Academy Award	15 22(62)/94 = 14.5	7 22(32)/94 = 7.5	22
Nobel Prize	29 40(62)/94 = 26.4	11 40(32)/94 = 13.6	40
Olympic gold medal	18 32(62)/94 = 21.1	14 32(32)/94 = 10.9	32
Total	62	32	94

Since all of the expected cell counts are at least 5, the second technical condition is met.

Note that these expected counts make the proportion who chose each achievement the same for the females as for the males (.234 Academy Award, .426 Nobel Prize, .340 Olympic gold medal).

26

H_0: There is no association between gender and choice of lifetime achievement.

H_a: There is an association between gender and choice of lifetime achievement.

The calculation for the chi-square test statistic uses all six cells:

$$\chi^2 = \frac{(15-14.5)^2}{14.5} + \frac{(7-7.5)^2}{7.5} + \frac{(29-26.4)^2}{26.4} + \frac{(11-13.6)^2}{13.6} + \frac{(18-21.1)^2}{21.1} + \frac{(14-10.9)^2}{10.9} = .01724 + .03333 + .25606 + .49706 + .45545 + .88165 = 2.14$$

The degrees of freedom of this statistic are $(3-1)(2-1) = 2$, and the corresponding chi-square distribution is shown here.

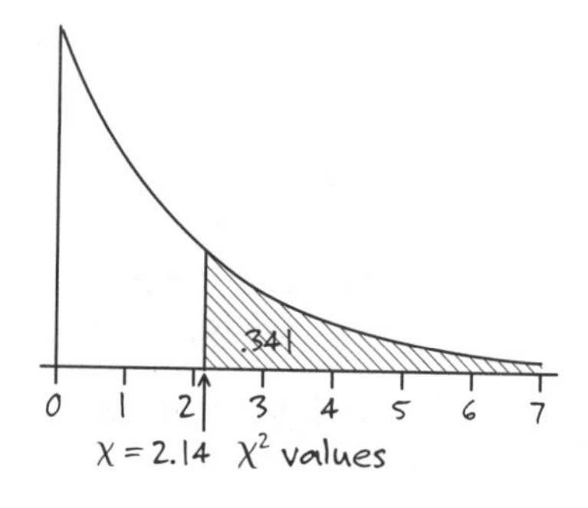

From *WS* Table IV you see that the corresponding p-value is larger than .2. (Technology indicates that it is .341.)

Since the p-value is larger than .05, you do not reject the null hypothesis at the 5% level. Thus, you do not have convincing evidence that response to the lifetime achievement question is associated with gender. In other words, the sample data do not provide compelling evidence that male and female statistics students at this university respond differently to this question.

Note that the largest discrepancy between the observed counts and the expected counts comes from the (Olympic gold medal, Males) cell. Looking at that cell, the observed count (14) is larger than the expected count (10.5). More males picked that response than you would have expected if males and females behaved similarly. This matches what you saw in the bar graph (with the additional interpretation that this is the biggest discrepancy). But the p-value of the test reveals that this discrepancy would not be surprising even if males and females in the population would respond the same way to the question.

KEY CONCEPTS

- A chi-square test enables you to test whether the association between two categorical variables is statistically significant. It can be applied when two categorical variables have been recorded for every observational unit in a random sample.
  - The hypotheses and conclusion of the test will be worded in terms of association or dependence between the two variables. (Make sure you identify both variables and your definition describes them as *variables*.)

- The technical conditions for the chi-square test should sound familiar—a random sample and a large sample size. With this test the sample size condition is checked by looking at the expected cell counts.
- The interpretation of the p-value is the same as always: if the null hypothesis is true, how often you would obtain a test statistic at least this extreme.
  - Large χ^2 values provide stronger evidence against the null hypothesis than small χ^2 values. Therefore you always want to decide whether the χ^2 statistic is larger than you would expect by chance. So, you will always calculate a *one-sided* p-value with chi-square procedures.
- There is no corresponding confidence interval formula for the chi-square test, because there is no simple parameter to be estimated. However, you can follow up a significant chi-square test by looking at the terms that comprise the chi-square sum and determining which terms are the largest. These correspond to the cells where the discrepancy between the observed counts and the expected counts is the greatest. You can assess the direction of the association by noting whether the observed count is larger than or smaller than the expected count in these cells.

CALCULATION HINTS

- Keep a few extra decimal places with your intermediate calculations and only round at the end.
- Remember to divide by the expected counts, not by the observed counts, when calculating the χ^2 statistic.
- Remember to square the numerator for each term in the sum.

COMMON OVERSIGHTS

- Checking the sample size technical condition by looking at the observed counts instead of the expected counts.
- Trying to use the chi-square test with quantitative data.
- Trying to compare the two column categories by averaging the counts in the two-way table, for example, averaging the three counts in the example to compare males and females. Remember to begin the analysis by asking yourself whether the variables of interest are quantitative or categorical and to only use the chi-square procedure with two categorical variables.
- Stating a "direction" in the alternative hypothesis. You can only state "there is an association" or "the variables are not independent." You can't test for positive or negative associations with the chi-square procedure.
- Using a table other than Table IV to locate the p-value.

WHAT WENT WRONG?

The following are examples of errors in the analysis of the example problem. In each case, explain what is wrong.

In a recent poll, researchers sampled 50 children, 50 teenagers, and 50 adults and determined whether each person knew all the words to "The Star-Spangled Banner." They found 22 children, 27 teenagers, and 20 adults that knew the words.

What is wrong with the following test statistic calculation?

$$\chi^2 = \frac{(22-23)^2}{23} + \frac{(27-23)^2}{23} + \frac{(20-23)^2}{23} = 1.13$$

Solution:

Only half the cells have been used. This calculation should involve six terms.

$$\chi^2 = \frac{(22-23)^2}{23} + \frac{(27-23)^2}{23} + \frac{(20-23)^2}{23} + \frac{(28-27)^2}{27} + \frac{(23-27)^2}{27} + \frac{(30-27)^2}{27} = 2.09$$

Topic 27:
INFERENCE FOR CORRELATION AND REGRESSION

In this topic you studied inference procedures for determining whether there is a relationship between two quantitative variables and for estimating the magnitude of the population correlation coefficient and population slope. Keep in mind that a thorough analysis of two quantitative variables begins with the numerical and graphical summaries you studied in *WS* Topics 8–11. You should be comfortable reading computer output for these analyses but should also consider how to verify the technical conditions of the inference procedures.

Example: A statistics professor collected data from the local newspaper on a sample of 25 homes sold in Bakersfield, CA, in April 2003. He wanted to know whether the size of the house (total square footage) would be a good predictor of its sale price in $1000's. The results are displayed in the following scatterplot.

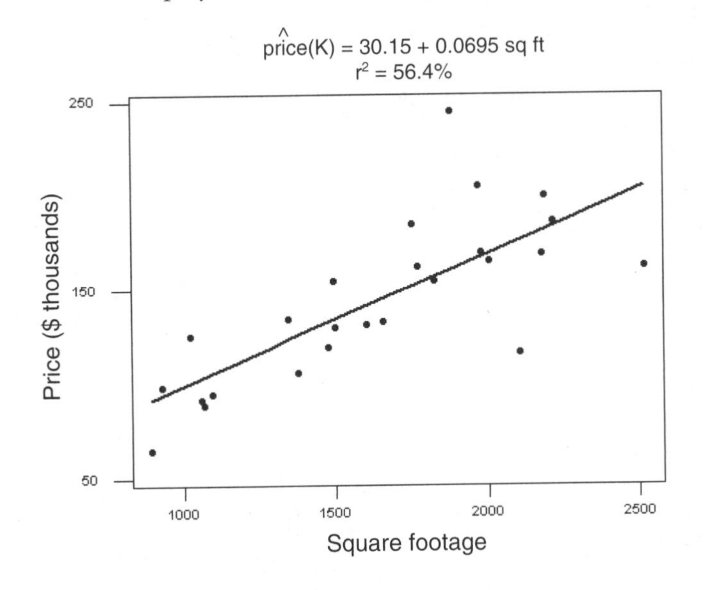

a. What is the value of the correlation coefficient for these data?

27

b. Carry out a test of significance to decide whether the correlation coefficient between the price and size for all houses in this population is positive.

c. What additional information do you need to carry out the details of a test of significance for the population slope coefficient? What does this number represent?

d. Based on the following Minitab® computer output, do you find significant evidence of a positive relationship between square footage and price in this population?

```
The regression equation is
price(K) = 30.1 + 0.0695 sq ft
```

Predictor	Coef	SE Coef	T	P
Constant	30.15	21.63	1.39	0.177
sq ft	0.06950	0.01275	5.45	0.000

e. Using this output, calculate and interpret a 95% confidence interval for the population slope.

f. Explain how you would check the technical conditions for the preceding procedure.

g. Based on the following residual plots, would you consider the technical conditions for the validity of the inference procedures about slope coefficient met?

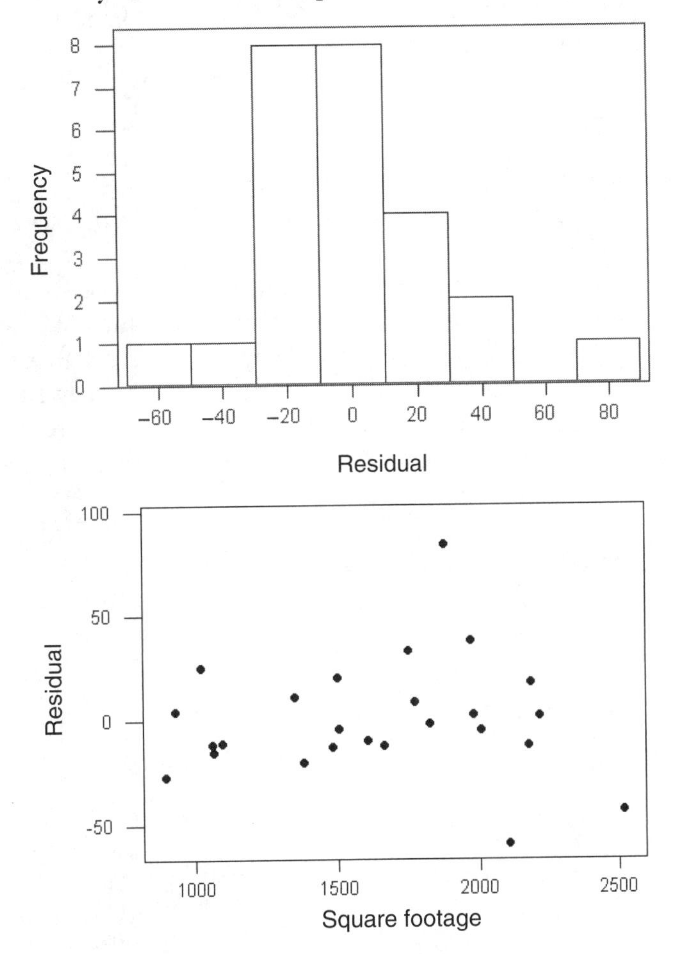

Solution:

a. The sample correlation coefficient is equal to the square root of r^2:

$$\sqrt{.564} = \pm .751.$$

Since the scatterplot shows the relationship to be positive, the correlation coefficient is $r = .751$.

b. Let ρ = the correlation coefficient between the prices and square footage for all houses in the population.

H_0: $\rho = 0$ (There is no linear association between price and square footage.)
H_a: $\rho > 0$ (There is a positive linear association between price and square footage.)

Next you check the technical conditions:

1. You were not told how the houses were selected. Presumably they were all the homes that sold that month, but you may be willing to consider them representative of all houses sold recently in Bakersfield, CA.
2. This second condition (normal distribution for both variables) is not met. The following histograms display the prices and the square footages. There is evidence that these distributions are skewed to the right, which you might expect based on the nature of real estate. For this reason, it is risky to perform and interpret the test statistic and p-value.

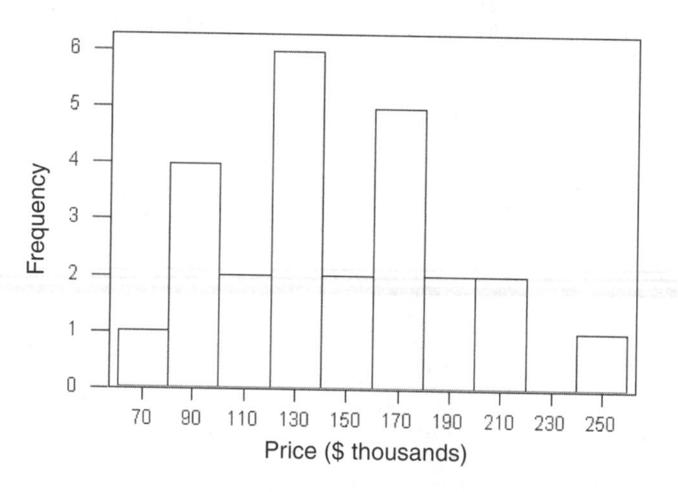

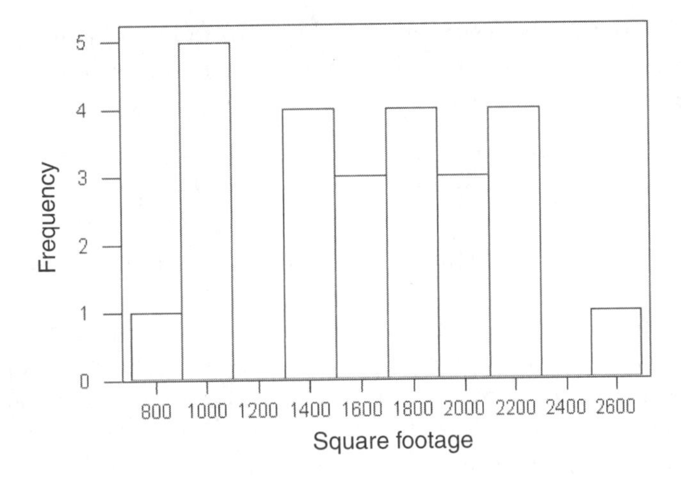

Note: If it had been appropriate to conduct this test, you would have found the following test statistic:

$$t = \frac{.751\sqrt{23}}{\sqrt{1 - .751^2}} = 5.45$$

The p-value is $\Pr(T_{23} > 5.45)$, which *WS* Table III reveals to be less than .0005. The conclusion would be that the sample data provide strong evidence of a positive association between house price and size in the population. But, again, you should draw this conclusion with some hestitancy because the technical conditions for this procedure are suspect.

c. To standardize the sample slope you would need the value of SE(b), the standard error of the sample slope coefficient. This value represents the amount of sampling variability in the slopes that would be obtained from taking different samples of 25 houses from this population of houses.

d. Using the row labeled "sq ft," you find the t-value and the p-value for the slope.

> H_0: $\beta = 0$ (There is no association between price and square footage in this population.)
> H_a: $\beta > 0$ (There is a positive association between price and square footage in this population.)

From the computer output, you find that $t = (.06950 - 0)/.01275 = 5.45$ and p-value = .000/2 (to three decimal places). Note that you need to divide the p-value reported by Minitab (and most other statistical software packages) by 2 because this software package assumes a two-sided alternative.

Notice that the preceding procedure is equivalent to what you did with the correlation coefficient test!

With such a small p-value (i.e., it is much less than .05), you can easily reject the null hypothesis. There is statistically significant evidence of a positive

27

relationship between price and square footage in this population. However, you can't use this analysis until you verify the technical conditions, which are slightly different than those for the test of the correlation coefficient.

e. A 95% confidence interval for β would use $t_{23} = 2.069$. If you plug the output values into the formula $b \pm (t^*_{n-2}) \times SE(b)$, you get

$$.06950 \pm 2.069(.01275) = .0695 \pm .0264 \text{ or } (.043, .096)$$

If the four technical conditions for this procedure are met, you are 95% confident that the slope coefficient to predict price from square footage for the population of homes for sale in Bakersfield is between .043 and .096 thousand dollars per square foot. In other words, you are 95% confident that the average increase in sale price for each additional square foot of size for houses sold in Bakersfield during early 2003 is between $43 and $96. However, you can't use this analysis until you verify the technical conditions.

f. To check the technical conditions, first determine whether the data can be considered a random sample from the population (which was discussed earlier). To check the other three conditions (linearity, normality of the response at each value of x, and equal standard deviations at each value of x), you must look at the residual plots:
(1) histogram of residuals (to check normality condition)
(2) residuals versus square footage (to check linearity and equal standard deviation).

g. The residual plots do not reveal marked problems:
- The distribution of the residuals is roughly symmetric.
- The standard deviation of the residuals appears to be roughly constant across the square footages except perhaps for a bit more variability due to two or three of the larger homes.

Since the technical conditions are not severely violated, you can proceed in interpreting the test statistic, p-value, and confidence interval calculated in parts d and e.

KEY CONCEPTS

- This topic completes the discussion you began in *WS* Topics 8–11. You should still begin your analysis of relationships between quantitative variables by examining graphical (scatterplot) and numerical (correlation coefficient) summaries and by considering outliers and influential observations.
- The techniques in this topic are appropriate when you have a random sample with two quantitative measurements for each observational unit. However, it is not always appropriate to carry out the inferential procedure, so make sure you examine the technical conditions first.
- If you do carry out the inference procedure, you can state the hypotheses and conclusions either in terms of the association or dependence between the variables or in terms of the population parameter (ρ or β). In these situations it is possible to test

for both one-sided and two-sided alternatives (as opposed to the chi-square procedures), so you must consider whether the p-value should be one-sided or two-sided.

- It is very important to be able to pick out the relevant numbers from computer output. Practice until you are comfortable doing so.

CALCULATION HINTS

- If you need to convert from r^2 to r, remember to consider whether r is positive or negative.
- When using computer output, make sure you use the test statistic and p-value corresponding to the slope rather than the intercept.
- When using computer output, make sure you determine whether the p-value reported is one-sided or two-sided.

COMMON OVERSIGHTS

- Not knowing which residual plots to examine or what to look for in them in checking the technical conditions.
- Assuming that a statistically significant relationship between two variables implies a cause-and-effect relationship.

WHAT WENT WRONG?

The following are examples of errors in the analysis of the example problem. In each case, explain what is wrong.

1. The sample slope coefficient reveals that a house's price goes up by $69.50 for each additional square foot of size.

2. If the technical conditions had been satisfied, concluding that the very small p-value would suggest that you should reject H_0.

3. If the technical conditions had been satisfied, the very small p-value would suggest that there is no linear association between a house's size and its price.

4. If the technical conditions had been satisfied, and if the p-value had been larger than .1, you would have concluded that the sample data provide strong evidence that there is no association between a house's size and its price.

5. Adding square footage to a house causes the price to increase by $69.50 on average.

Solutions:

1. The slope is .695, and the response variable is "thousands of dollars," so this is close to a correct interpretation, but it does not take the variability of the prices about the regression line into account. A better interpretation would be to say that the price of a house increases by *an average of* $69.50 for each additional square foot of size.
2. This is correct but incomplete. Never leave a conclusion at "reject H_0." Remember to *always* relate your conclusion back to the context.
3. The very small p-value would suggest strong evidence against the null hypothesis. But the null hypothesis says that the population slope equals zero, which means that there is *no* linear association between house price and size in the population. So, this conclusion is backward: The small p-value actually provides strong evidence that there *is* a linear association between house price and size in the population.
4. This conclusion commits the error of "accepting H_0." Remember that a test of significance assesses the strength of evidence *against* the null hypothesis, *not in favor* of it. A large p-value would suggest no evidence against the null hypothesis of no association, but it would not suggest strong evidence of no association.
5. This conclusion states that the small p-value implies a causal relationship. Since this was an observational study, such a conclusion is not valid.

FURTHER EXPLORATION

As we said following *TK* Topic 22, once you have learned several procedures, it is important to practice distinguishing which procedure applies in a given situation. Now that you have learned even more procedures, this task is all the more important to practice and become comfortable with. A good first step is always to identify the observational units and variables, and then determine the number of variables and the classification of each variable's type (categorical or quantitative). Below is a list of the procedures you have encountered.

1. One-sample proportion z-test
2. One-sample proportion z-interval
3. One-sample mean t-test
4. One-sample mean t-interval
5. Two-sample proportions z-test
6. Two-sample proportions z-interval
7. Two-sample means t-test
8. Two-sample means t-interval
9. Chi-square test
10. t-test for regression slope, correlation
11. t-interval for regression slope, correlation
12. None of these procedures is appropriate.

For the following situations, state what procedure you would use. If the choice of procedure is unclear from the information given, state what additional information you would need. If you choose a test of significance, state the null and alternative hypotheses. If you choose a confidence interval, define the relevant population parameter(s).

a. A campus administrator wants to know whether some campus groups are more likely to consume alcohol than others. He takes a random sample of 1500 students and classifies them as high-risk or low-risk drinkers and whether they belong to a sorority, fraternity, or neither.

b. A researcher wants to determine whether people with "positive attitudes" tend to live longer than those without positive attitudes. She collects data on those who were classified with a positive attitude and those who were not, and records how long they lived.

c. A student project group wants to know whether the color of a person's car is related to how fast he or she drives. They time how long it takes cars to travel between two points on the local highway and classify the cars as "racy colored" (red, black), "light" (white, tan, silver), and "other" (any other color).

d. A university is trying to determine whether parking is a problem on its campus. The student newspaper contacts a random sample of 200 students and asks whether they are frustrated with the parking situation. They want to estimate the proportion of students at the college who are frustrated with the parking situation.

e. A student reporter wants to know whether Democrats at the school were more likely to vote for candidate NW than were Republicans in the last election. He takes random samples of students belonging to the Democratic and Republican parties and asks whether they voted for candidate NW.

f. A psychologist wants to know whether the amount of time a couple cohabits before marriage is related to how long the marriage lasts. She selects a random sample of 500 divorced couples and for those who cohabited before marriage she records how many months they lived together before they were married and then the number of years they were married before they divorced.

g. The Best Buy electronics store wants to estimate how much more men tend to spend in the store than women. The marketing department collects receipts for a random sample of 100 customers and examines the total amount of the bill.

Solution:

a. You would use a chi-square test.
Observational units = students
Variable 1 = whether or not the student belongs to a sorority, fraternity, or neither (categorical)
Variable 2 = whether the student is a high-risk or low-risk drinker (categorical)
[*Note*: Since variable 1 has three categories, you can't use a two-sample procedure.]
H_0: There is no association between Greek membership and risk level.
H_a: There is an association.

b. You would use a two-sample means *t*-test *as long as the people were randomly selected.*
Observational units = people
Variable 1 = whether or not the individual has a positive attitude (categorical)
Variable 2 = number of years lived (quantitative)
Let μ_1 = mean lifetime of those with a positive attitude.
Let μ_2 = mean lifetime of those without a positive attitude.
H_0: $\mu_1 = \mu_2$ (There is no difference in mean lifetime between these two populations.)
H_a: $\mu_1 > \mu_2$ (Those with positive attitudes tend live longer on average.)

c. This study attempts to compare three means; none of these procedures do that.
Observational units = cars
Variable 1 = speed (quantitative)
Variable 2 = color classification (categorical)
H_0: $\mu_1 = \mu_2 = \mu_3$ (The average speed is the same for each color population.)

d. You would use a one-sample proportion *z*-interval.
Observational units = students
Variable = whether or not the student is frustrated with the parking situation (categorical)
Let θ = the proportion of all students at this university who are frustrated with the parking situation.
You want to find a confidence interval for θ.

e. You would use a two-sample proportions *z*-test.
Observational units = students
Variable 1 = whether the students classify themselves as Democrats or Republicans (categorical)
Variable 2 = whether or not the students voted for candidate NW (categorical)
[*Note*: Assuming you have independent samples and since you want to perform a one-sided test, the two-sample proportions test is more appropriate than the chi-square test.]
Let θ_1 = the proportion of Democrats at this school who voted for candidate NW.
Let θ_2 = the proportion of Republicans at this school who voted for candidate NW.
H_0: $\theta_1 = \theta_2$ (Republicans and Democrats were equally likely to vote for candidate NW.)
H_a: $\theta_1 > \theta_2$ (Democrats at this school were more likely to vote for candidate NW than Republicans.)

f. You would use a *t*-test for regression slope or correlation.
Observational units = couples
Variable 1 = length of time the couple cohabited before marriage (quantitative)
Variable 2 = number of years the couple's marriage lasted (quantitative)
H_0: There is no association between length of cohabitation and length of marriage.
H_a: There is an association between length of cohabitation and length of marriage.
If β is the population slope these hypotheses can also be written as
H_0: $\beta = 0$
H_a: $\beta \neq 0$.

g. You would use a two-sample means *t*-interval.
Observational units = customers
Variable 1 = gender (categorical)
Variable 2 = total amount of bill (quantitative)
Let μ_1 = the average amount spent by men.
Let μ_2 = the average amount spent by women.
You want to find a confidence interval for $\mu_1 - \mu_2$.
If ρ is the population correlation coefficient, these hypotheses can also be written as
H_0: $\rho = 0$
H_a: $\rho \neq 0$.

Index

A

alternative hypothesis, 166, 167, 175
association
  cause vs., 49
  graphical displays of, 43–46

B

bar graphs, 4–5, 37, 39
bias, 73, 74, 75, 77, 85
binary variables, 4, 184
bivariate distributions, 39
blind studies, 84
blocking, 83, 84, 85
boxplots, 22, 25, 31, 32

C

categorical variables
  chi-square test and, 214
  comparing distributions of, 35–42
  identification of, 5–6
  sampling distrubutions and, 131
cause-and-effect relationships, 37–38, 45–46, 49, 199
  study design and, 69, 84–85
center, 12, 13, 17–20
Central Limit Theorem, 110, 112–113, 131–139, 164–165
chi-square test, 211–216
conditional distributions, 37–42
conditional percentages, 39
conditions. *See* technical conditions
confidence intervals
  comparison for means, 206, 207, 208
  comparison for proportions, 198, 199, 200
  considerations about, 190–191
  means, 153–156, 159–160
  proportions, 143–152
  for slope coefficient, 221
confidence level, 147, 156
confounding, 84, 85, 86
correlation coefficient (*r*), 47–49, 56, 217–221, 222–223
count, vs. rate or proportion, 8–9
critical values, 180

D

data, good, 69
degrees of freedom, 155, 156, 180, 205, 207
dependent variables, 37
digits of accuracy, 24
distributions
  bivariate, 39
  comparing, 29–33, 35–42
  conditional, 37–42
  describing and displaying, 11–16
  normal, 97–105
  outliers of. *See* outliers
  sampling. *See* sampling distributions
  skewed, 13
dotplots, 5

E

empirical probability, 93
empirical rule, 24, 101, 111
empirical sampling distribution, 116
equal likeliness, 93, 96
even distribution, 14
experiments, 82, 84
explanatory variables, 36, 39, 55

F

fitted value, 56
five-number summary, 24, 25

G

granularity, 13
graphs
  axes, plotting variables on, 13, 45
  bar graphs, 4–5, 37, 39
  boxplots, 22, 25, 31, 32
  choosing type of, 13
  dotplots, 5
  histograms, 13
  scatterplots, 45–46
  segmented bar graphs, 37, 39
  stemplots, 13, 16, 31

H

histograms, 13
horizontal axis, 13, 45

I

independent variables, 37, 41
inference, 69, 145
influential observations, 60–61, 63–64
interquartile range (IQR), 21–26, 32

L

least squares regression
  determining line, 51–59
  influential observations and, 60–61, 63–64
  slope of regression lines, 54, 55, 59, 217, 221–224

M

margin of error, 146
marginal percentages, 39
mean
  comparison of two, 203–210
  confidence intervals, 153–156, 159–160, 206, 207, 208
  as measure of center, 17–20
  of sample, 121, 124–125
  tests of significance, 177–181, 184–185
measures of center, 17–20
measures of spread. *See* spread
median, 17–20
mode, 19
modified boxplots, 31, 32

N

normal distributions, 97–105
null hypothesis
  chi square test and, 214, 215
  comparisons and, 199–200, 207
  and population slope, 224
  tests of significance, means, 179, 181, 184, 185
  tests of significance, proportions, 165, 166, 167–168, 170
  type I errors and, 190
  type II errors and, 190, 191

O

observational units, 3–6
one-sided alternatives, 184, 200
outliers
  regression and, 55
  removing from analysis, 12–13
  resistance to, 18, 21, 49
  rule for determining, 32, 33

P

p-value
- chi-square test and, 214, 215
- comparisons and, 198, 199, 205, 206, 209
- null hypothesis and, 166, 167, 168, 184–185, 190
- tests of significance, means, 178–179, 180, 181, 185
- tests of significance, proportions, 165–166, 167, 168–170, 175
- variability and, 174

parameter, 73, 74, 111, 191
percentage, proportion vs., 9
placebo effect, 83, 84
population, 72, 73, 74, 75
- population size, 74, 112

population distribution, 125
precision, 75
probability, 91–96
proportion
- comparison of two, 195–200
- confidence intervals, 143–152, 198, 199, 200
- percentage compared to, 9
- tests of significance, 161–170, 175

Q

quantitative variables
- comparing distributions of, 29–33
- correlation coefficient for, 47–49, 56, 217–221, 222–223
- graphical display of, 43–46
- identification of, 5–6
- sampling distributions of, 131

quartiles, 24
question wording and order, 73

R

r value. *See* correlation coefficient
r^2 value, 56, 59
random assignment, 81–88
random number table, 74
random sampling, 74–75, 84–85
randomized comparative experiments, 81–88
range, 21, 24
ratios, 8–9
regression line. *See* least squares regression
regression slope, 217, 221–224
relative position, 101
residual plots, 64–66, 221–222
residuals, 56, 64
resistance to outliers, 18, 21, 49
response variables, 36, 55
rounding, 56, 156, 180

S

sample, 72, 73, 74, 75, 125
sample distribution, 125
sample size
- bias not reduced by, 74
- confidence intervals and, 154, 155, 191
- vs. number of samples, 112
- probability and, 93–94
- sampling distribution and, 111, 112
- tests of significance and, 175
- variability and, 116, 131, 136

sampling, 69–80
- bias and, 73, 74, 75, 77, 85
- designing studies, 81–88

sampling distributions
- defined, 125
- of mean, 121, 124–125
- of proportion, 107, 111–113, 116–119
- type of variable and, 131

sampling frame, 73, 74, 75
sampling variability, 74, 110–111, 116, 130–131, 135–136, 146, 174, 221
scatterplots, 45–46
segmented bar graphs, 37, 39
shape, 12, 13
short-run deviations, 93, 94
significance. *See* tests of significance
simple random sample, 74–75, 84–85
Simpson's paradox, 38
skewness, 13
slope coefficient, 217, 221–224
spread. *See* variability
standard deviation, 21–26, 32
standard error, 199
standard score. *See* *z*-score
statistic, 73, 74, 111
stemplots, 13, 16, 31
strength of associations, 49
studies, designing, 81–88
symmetric, distribution, 13

T

*t**, 180
t-distribution, 155, 180
t-interval, 154
t-test, 180, 181, 205
t-value, 156, 158
table of random numbers, 74
technical conditions
- chi-square test and, 215
- comparison of means and, 206, 207
- comparison of proportions and, 199
- and confidence intervals, means, 154, 155, 156, 160
- and confidence intervals, proportions, 145, 146, 148
- correlation coefficient and, 220–221
- regression and, 221–222
- and tests of significance, means, 178, 179, 184, 185, 186
- and tests of significance, proportions, 165, 166, 169, 170

tendency, 31, 45
test statistic, 166–168, 169, 178–179, 180
tests of significance
- chi-square, 211–216
- comparison of means, 206–208
- comparison of proportions, 198–200
- considerations about, 184–185, 190–191
- means, 177–181, 184–185
- proportions, 161–170, 175
- regression, 217, 221–224

theoretical probability, 93
transformation of variables, 65
two-sided alternatives, 184–185, 200
two-way tables, 39, 211–216
type I and type II errors, 190, 191

U

unit of measurement, specifying, 13

V

variability (spread)
- anticipation of, 18
- bumpiness distinguished from, 24
- confidence intervals and, 155, 160
- description of, 12, 13
- measures of, 21–26
- p-value and, 174
- sample size and, 116, 131, 136
- sampling, 111, 116, 124, 130, 131, 135, 136, 146

variables
- appropriate, 7–9
- binary, 4, 184
- conditioning on, 38–39
- dependent, 37
- explanatory, 36, 39, 55
- horizontal axis and, 13
- identification of, 3–6
- independent, 37, 41
- response, 36, 55
- transformation of, 65
- *See also* categorical variables; quantitative variables

vertical axis, 13, 45

Y

y-intercept, 55, 56

Z

*z**, 180
z-score (standardized score), 24, 99–100, 101–102, 104–105